T0342180

**The Load-Pull Method of RF and Microwave
Power Amplifier Design**

The Load-Pull Method of RF and Microwave Power Amplifier Design

John F. Sevic, Ph.D.

Registered Office
John Wiley & Sons, Inc., 111 River Street, Hoboken, NJ 07030, USA

Editorial Office
111 River Street, Hoboken, NJ 07030, USA

For details of our global editorial offices, customer services, and more information about Wiley products visit us at www.wiley.com.

Wiley also publishes its books in a variety of electronic formats and by print-on-demand. Some content that appears in standard print versions of this book may not be available in other formats.

Library of Congress Cataloging-in-Publication Data

Names: Sevic, John F., author.
Title: The load-pull method of RF and microwave power amplifier design /
   John F. Sevic, Vice President of Millimeter-Wave Engineering, Maja Systems,
   Milpitas, CA.
Description: First edition. | Hoboken, NJ : John Wiley & Sons, Inc., 2020.
   | Includes index.
Identifiers: LCCN 2020009983 (print) | LCCN 2020009984 (ebook) | ISBN
   9781118898178 (hardback) | ISBN 9781119078067 (adobe pdf) | ISBN
   9781119078036 (epub)
Subjects: LCSH: Power amplifiers–Design and construction. | Amplifiers,
   Radio frequency–Design and construction. | Microwave amplifiers–Design
   and construction.
Classification: LCC TK7871.58.P6 S45 2020 (print) | LCC TK7871.58.P6
   (ebook) | DDC 621.3841/2–dc23
LC record available at https://lccn.loc.gov/2020009983
LC ebook record available at https://lccn.loc.gov/2020009984

Cover Design: Wiley
Cover Image: © Elesey/Shutterstock

Set in 9.5/12.5pt STIXTwoText by SPi Global, Chennai, India

10 9 8 7 6 5 4 3 2 1

This book is dedicated to the next generation of load-pull experts: may your experiences be as enriching and rewarding.

Contents

List of Figures

List of Tables

Acronyms, Abbreviations, and Notation

ADC	analog-to-digital converter
APC	amphenol precision connector
BJT	bipolar junction transistor
BW	bandwidth
CAD	computer-aided design
CDMA	code division multiple access
CMOS	complementary metal oxide semiconductor
CW	continuous wave
DCS	digital cellular system
DIN	deutsche internationale Normen
DUT	device under test (used synonymously with transistor)
EDA	engineering design automation
EER	envelope elimination and restoration
GSM	global system for mobile communication
HF	high frequency
LP	load-pull
LTE	long term evolution (4G)
MMIC	monolithic microwave integrated circuit
MOS	metal oxide semiconductor
PCS	personal communication system
PN	pseudo noise
SA	spectrum analyzer
SOL	short-open-load
TE	transverse electric
TEM	transverse electromagnetic
TRL	thru-reflect-line
VHF	very high frequency
VNA	vector network analyzer
VSWR	voltage standing wave ratio

WCDMA	wideband CDMA (3G)
RMS	root mean square
PEP	peak envelope power
PAE	power-added efficiency (%)
PAR	peak-to-average ratio (dB)
CCDF	complex cumulative distribution function
VBW	video bandwidth (a measure of instantaneous modulation BW)
ω_0	resonant radian frequency (rad/s)
BW_3	3 dB bandwidth of frequency response magnitude (Hz)
MCPA	multi-carrier power amplifier
DPD	digital pre-distortion
HBT	heterojunction bipolar transistor
LDMOS	laterally diffused metal oxide semiconductor
HEMT	high electron mobility transistor
GaN	gallium nitride
GaAs	gallium arsenide
ACPR	adjacent channel power ratio (usually associated with 2G CDMA)
ACLR	adjacent channel leakage ratio (usually associated with 3G WCDMA and 4G LTE)
EVM	error vector magnitude
CDP	code-domain power
IM_3	third-order intermodulation ratio (dB)
AM–AM	amplitude to amplitude conversion distortion (gain compression)
AM–PM	amplitude to phase conversion
ET	envelope tracking
EF	envelope following
T	matching network impedance transformation ratio
T_p	L-section prototype impedance transformation ratio
IL	insertion loss (dB)
ESR	equivalent series resistance
CW	clockwise rotation (on the Smith chart toward the source)
PVT	process, voltage, and temperature (an acronym for these variables associated with robustness)
BOM	bill of materials
PDK	process development kit

Preface

When I was asked to build a handset power amplifier (PA) design group at Qualcomm in San Diego, many years ago, it became evident that comprehensive treatment of a systematic RF PA design process in a concise self-contained volume was necessary. At the time, the most common design process was trial and error using triple-stub tuners or, often times, with the ubiquitous roll of copper tape and X-Acto knife. Automated load-pull was still a relatively new tool, suffering from several limitations, including harmonic control, limited impedance range, and linearity impairment. Indeed, first-generation automated load-pull was incapable of creating the impedance necessary for a handset code division multiple access (CDMA) PA. It was not entirely plausible that load-pull would evolve into the universal standard of RF PA design that today it currently is, 20 years later.

Writing a book on load-pull for RF PA design thus posed somewhat of a risk, due to its well-known limitations and skepticism by a technically adept audience, often highly critical and guarded of new design methods. That load-pull has evolved into the most popular method for systematic RF PA design is precise because all of the early limitations of load-pull have been rigorously addressed in the intervening time since Qualcomm. The timing is therefore right for the present volume, containing in a single sitting comprehensive treatment of the load-pull method for RF PA design.

The popularity of the load-pull method, to the exclusion of all others, is because absolute performance is systematically, accurately, and rapidly appraised without resorting to slow and uncertain trial and error or reliance on abstruse and nonphysical mathematics. Because the load-pull method distinguishes between management of nonlinearity versus analysis of nonlinearity, it transforms what is essentially an intractable nonlinear mathematical problem into a series of variable-impedance measurements followed by utilization of well-understood linear network theory for matching network design. The uptake of the load-pull method can also be attributed to its accessibility by a broad class of new electrical engineers, most of who successfully design an RF PA on their first attempt.

A distinguishing feature of the present volume is an attempt to instill in the reader a strong sense of empirical judgment, which I believe is central to success as an RF PA engineer. During my 20-year career designing RF PA's, I have observed an affinity by a surprising number of engineers, and not a few academics, an overreliance on intractable mathematical analysis for RF PA design. Do not get me wrong: I embrace mathematics when necessary. But for the load-pull method presented to the reader of this book, mathematics beyond freshman calculus is unnecessary. Instead of reducing the physical world to a few oversimplified, yet intractable equations, I prefer instead to imbue a sense of judgment to infer reality and implement a robust design based on sound engineering principles that achieves first-pass success. The great Silicon Valley RF engineering educator and icon, Dr. Frederick Terman, well expresses the theme of my book [1]. Paraphrasing from Fred's classic treatise on radio engineering, he observes that...

...equations can only go so far. At some point, a little bit of hand-tuning with a screwdriver becomes absolutely essential for successful RF PA design.

I wholeheartedly agree with Fred, and make no apologies about the absence of complex and intractable mathematics and in its place deliberate emphasis on empiricism and graphical methods. It is suspected the reader will agree with, and appreciate, this approach.

The book is written to be accessible by practicing RF engineers with exposure to intermediate distributed-parameter networks, intermediate analog circuits, and introductory microwave semiconductor physics typically associated with the BSEE or equivalent. The well-prepared reader will be comfortable with the material in the Gonzales text on microwave amplifier design, particularly its excellent treatment of Smith chart operations, and the Raab text on solid-state radio engineering, while the advanced reader will be familiar the Maas text on nonlinear microwave circuit design [2–4]. The two Cripps books on RF PA design are highly complementary companions to the load-pull method of RF PA design treated here, and it is recommended the reader be comfortable with each of these texts [5, 6]. In addition to being suitable for the practicing RF PA engineer, undergraduate and graduate students will benefit from the book, as will those attending industry classes such as those offered by Besser and Associates. As an RF engineer, I have benefited from Besser courses, and highly recommend them.

Although the PA can be found in applications spanning high frequency (HF) to sub-millimeter-wave, emphasis in the book is on applications for wireless communication systems. The load-pull method of PA design presented is general and can be applied to other bands or applications, though the bulk of the readership is expected to be engaged in contemporary wireless standards, such as long term evolution (4G) (LTE) and 5G, using such architectures as Doherty, envelope tracking (ET), and digital pre-distortion (DPD). Each of the chapters is written to be used independently of each other, although it is recommended to be familiar with the vector network analyzer (VNA) and its calibration, because of its central importance to the practice of load-pull.

Unique to the book is comprehensive treatment of load-pull accuracy analysis and verification, including introduction to the now-universal ΔG_T method [7, 8]. A rigorous introduction to matching network design using load-pull data is provided, along with several worked examples for both lumped and distributed matching networks.

In parallel with the growth of load-pull has been a concomitant rise in the use of engineering design automation (EDA) tools for RF design, commercial tools based on the harmonic-balance method and the circuit-envelope method. While it is possible to use EDA exclusively for RF PA design, with additional allowance for post-fabrication tuning, I instead take the agnostic position that EDA and load-pull are not substitutes but rather compliments. To this theme specifically, the method described here can in its entirety be used with EDA-based load-pull. For example, the Keysight Advanced Design System (ADS) has templates for load-pull simulation.[1]

A distinguishing feature of science is the ability of successive generations to build on the work of others, to constantly improve, adapt, and invent. To this point, I owe a debt of gratitude to many people, some of who were early inspirations and have now become close and trusted friends, including Dr. Fred Raab, Nathan Sokal, Dr. Steve Cripps, and Dr. Les Besser. My generation of RF PA engineers owe to these men a great debt in advancing the art and science of systematically rigorous RF PA analysis, design, and simulation.

I would also like to acknowledge Jim Long of Motorola Cellular Infrastructure Group, who hired me in 1990 as a fresh-out RF PA designer from Michigan Tech. Motorola was fortunate to have some of the best RF engineers in the world, and I would like to thank the many outstanding people I learned from there, especially Jim Long, Joe Staudinger, Mike Majerus, Dr. Mike Golio, Dr. Robert Stengel, Alan Wood, Roy Hejhall, and Helge Granberg.

I have been extremely fortunate to work with arguably two of the most important men in the modern load-pull business, Dr. Christos Tsironis, President of Focus Microwaves, and Gary Simpson, CTO of Maury Microwave. Their friendship and leadership is happily acknowledged.

---

1 These templates were first made by the author in HP Microwave Design System, the precursor to ADS, including the template for adjacent channel power ratio (ACPR) harmonic-balance and circuit-envelope simulation.

Many thanks are due to the Wiley staff, especially my Editor Brett Kurzman, for their patience and support during the writing of this book.

2020

John F. Sevic, Ph.D.
Los Gatos, CA

References

1 Terman, F. (1940). *Radio Engineering*. McGraw-Hill.
2 Gonzales, G. (1984). *Microwave Transistor Amplifier Design*. New York: McGraw-Hill.
3 Krauss, Herbert L., Bostian, Charles W. and Raab, Frederick H. (1980). *Solid-State Radio Engineering*. New York: McGraw-Hill.
4 Maas, S.A. (1988). *Analysis of Nonlinear Microwave Circuits*. London: Artech House.
5 Cripps, S.C. (2002). *RF Power Amplifiers for Wireless Communications*, 2e. London: Artech House.
6 Cripps, S.C. (2008). *Advanced Techniques in RF Power Amplifier Design*. London: Artech House.
7 Sevic, J.F. and Burger, K. (1996). Rigorous error analysis for low impedance load-pull, from Internal Notes at Qualcomm.
8 Sevic, J.F. (2001). *Theory of High-Power Load-Pull Characterization for RF and Microwave Transistors*, Chapter 7. Boca Raton, FL: CRC Press.

Foreword

Every engineer should understand the basic theory behind the circuit being designed. If you don't know the theory, you won't know what to expect in the lab. Often, relatively simple theory is sufficient to get a design "into the ballpark."

The utility of pure theory is often limited by the complexity of the circuit and the model of the transistor. Simulation offers a means of dealing with circuits and models that are too complex to be tractable with analytical mathematics. Consequently, simulation is a useful tool for further analysis and optimization of an RF-power circuit.

The accuracy of the simulation is of course dependent upon the accuracy and completeness of the models of the transistor and circuit. Real circuits include "stray" inductance and capacitance as well as nonzero lengths of interconnecting leads, and these are often difficult to determine with precision. A model for a given transistor is not always available. Simulators are expensive and often beyond the reach of small companies and individuals. The models of the transistors are often based upon operation the linear and mildly nonlinear regions, as these applications account for the majority of the sales. Significantly nonlinear operation such as switching modes may produce bizarre results. It is generally a good idea to compare the results of simulation to the performance of a real amplifier before relying too heavily upon the simulation.

Load-pull, in contrast to theory and simulation, requires neither knowledge of the circuit nor the a model of the transistor, nor even knowing the class of operation. One simply places the transistor into a basic circuit and connects its output to the load-pull tuner. The tuner then varies the load impedance over a specified range, after which contours of power output, efficiency, distortion, etc. are plotted on a Smith chart. I frequently extract a power versus efficiency curve that I then use it to select the operating point. Data can also be processed to produce just about any other trade-off of interest. The engineer selects a load impedance that gives the preferred combination of performance parameters and then proceeds to design the matching network.

The relative ease with which one can obtain useful results has made load-pull a popular technique. While the concept is simple, successful use requires knowledge of its limitations, accurate calibration, and the like. Dr. Sevic's book addresses this need. John is very well qualified to present this material as he has not only worked for both of the major manufacturers of load-pull tuners but also used load-pull himself in the design of RF power amplifiers.

Chapter 1 introduces the basic concepts, with some historical perspective. Chapter 2 then presents the passive and active methods of creating the impedances, along with advantages and limitations of each. Also addressed are how to incorporate harmonic impedances and how to produce very low impedances.

Chapter 3 deals with hardware issues. It begins with system architectures and block diagrams and proceeds to address generation of the test signals and measurement of the parameters of the output. Data processing and control strategies are addressed in Chapter 4. This includes strategies for mapping not just power and efficiency, but almost any conceivable performance parameter.

The ultimate goal of load-pull is usually to determine the load impedance that gives a desirable combination of performance parameters. Chapter 5 therefore addresses determining this impedance for different conditions and goals. Finally, Chapter 6 shows how to use load-pull data to design output matching networks. This includes various frequency responses and also the inclusion of harmonic impedances.

2020

Frederick H. Raab, Ph.D.
Green Mountain Radio Research
Boone, IA

Biography

John F. Sevic has over 30 years of experience in microwave engineering, from transistor design and modeling to advanced power amplifier design using stochastic optimization. Dr. Sevic began his career at Motorola, where he was involved with Si BJT and LDMOS characterization using load-pull and RF power amplifier design for wireless applications. At Qualcomm, Sevic invented the now ubiquitous long-term stochastic PA efficiency optimization concept for CDMA; this idea has been applied to virtually all wireless standards since 2000 and formed the basis of several PA start-ups. While at Cree and Tropian, Sevic developed the universal ΔG_T load-pull verification metric, TRL-based sub-1 Ω load-pull prematching deembedding theory, and real-time IR load-pull for LDMOS electrothermal model extraction. Sevic later worked at both Maury Microwave and Focus Microwaves, working on several advanced characterization tools, including nonlinear vector network analysis and prematching tuners. Currently, Sevic is Vice President of Engineering at Maja Systems, where he designs mmWave antennas.

Sevic has been involved with load-pull specifically for over 20 years, including developing the first ACPR-capable load-pull system, the first automated envelope baseband load-pull system, development of systematic two-tier TRL fixture deembedding, and the first IR real-time thermal imaging load-pull system. Sevic has extensive experience in high-power RF modeling, including development of self-consistent electrothermal models for LDMOS and HBT technologies, as well as significant experience in designing BJT, HBT, HEMT, and LDMOS discrete power transistors and MMIC's, mobile power amplifiers, and infrastructure power amplifiers.

Dr. Sevic is a Senior Member of the IEEE and served on the IEEE Microwave Theory and Techniques Editorial Review Board from 1996 to till date, the IEEE International Microwave Symposium TPC from 1994 to 2008, and the IEEE Automatic RF Techniques Group TPC from 1998 to 2005. Sevic received the best paper award at the 1997 ARFTG conference for describing the highest VSWR prematching fixture ever developed, with a record 500:1 VSWR.

Sevic has over 50 peer-reviewed publications, is lead inventor on 10 patents, with 5 pending, author of 3 book chapters, and author of the Wiley textbook The Load-Pull Method of RF and Microwave Power Amplifier Design. He received a BSEE from Michigan Technological University, and an MSEE from Illinois Institute of Technology, and a PhD in solid-state physics from University of California at Santa Cruz.

1

Historical Methods of RF Power Amplifier Design

1.1 The RF Power Amplifier

The RF power amplifier (PA) is a fundamental component of anything communicating wirelessly, being found in the ever ubiquitous mobile phone, the laptop, pad and tablet computers, as well as being found in many other applications, such as medical diagnostic tools, RF heating equipment, and navigation systems. Compared to adjacent sections of the radio, such as the receiver, and digital signal processing functions, such as the baseband modem, the RF PA can draw substantially more power, and thus is exposed to increased scrutiny due to its impact on battery life or base-station operating expense. The RF power transistor represents a significant cost factor because of die size, nature of the semiconductor material, and package cost, particularly for wireless infrastructure applications. Application-specific constraints also exist, particularly in 4G and 5G wireless systems, due to the time-varying envelope whose fidelity must be maintained, mandated by strict air-interface standards for spectral emissions.

Defining an RF PA is usually arbitrary, and is almost always relative. For example, the 2 W provided by a mobile RF PA might be considered high power compared to the CMOS transceiver IC driving it, whereas the 1 kW infrastructure RF PA may also be considered high power. Nevertheless, in contrast to the small-signal amplifier, the RF PA is unambiguously and uniquely distinguished by three key properties.

The RF PA operates in the large-signal regime where the voltage and current traverse the entire active region of the transistor, instantaneously probing, and sometimes dwelling in, cut-off, and saturation. In contrast, the small-signal amplifier operates in the weakly nonlinear regime, exhibiting limited instantaneous excursions over the load-line, with negligible harmonic content. Because of this, superposition holds for the small-signal amplifier, and its optimum terminating

The Load-Pull Method of RF and Microwave Power Amplifier Design, First Edition. John F. Sevic.
© 2020 John Wiley & Sons, Inc. Published 2020 by John Wiley & Sons, Inc.

impedances and performance over the entire Smith chart are uniquely and accurately described by its four *s*-parameters alone.

In contrast, because superposition does not hold over the large-signal operating regime of the RF PA, extrapolation over the entire Smith chart is not possible. The strong nonlinearities exhibited by the RF PA leads to the phenomenon of mixing products, a superset of intermodulation distortion, whereby spectral content at integer multiples of the stimulus appears at both the input and output. How these individual mixing products are terminated establishes the performance of the RF PA, particularly its linearity and efficiency, although these cannot be *a priori* known. RF PA design requires that measurements be made within a predefined region of the Smith chart to uniquely identify optimum terminating impedances and performance, but only in the area bounded by predefined region, since superposition does not hold.

At RF and microwave frequencies, the wavelength of the stimulus is on the order of the size of the physical dimensions of the network embedding RF power transistor die or package. This distributed-parameter environment, coupled with the extremely low impedance required by modern high-power RF power transistors, poses a challenge for synthesis and design of matching networks. To establish optimum performance in this environment, composed of multiple parasitic resonances, frequency-dependent losses, and distributed effects, considerable time-intensive experimentation is necessary to identify the optimum terminations that establish absolute performance. This is followed by design of a suitable matching network replicating the required impedance terminations at the relevant mixing product frequencies, usually at the harmonic and baseband frequencies. In contrast, small-signal amplifier design is only concerned with *s*-parameters at the fundamental frequency. This uncertainty, and need to present a specific impedance at many discrete frequencies, means matching network design for the RF PA involves multiple iterations and extensive trial-and-error design.

Management of heat, and its consequences, is one of the major challenges RF power transistor and PA designers face. Heat results in reduced gain, reduced effective power density, and impaired linearity. Moreover, each of these manifestations of heat generation is accentuated when the wavelength of the source signal is on the order of the physical dimensions of the die, as is often the case for high-power applications. In addition to the aforementioned effects on performance, reliability can be impaired or catastrophic failure can be induced due to localized self-heating, particularly under mismatch conditions, as commonly encountered in the handset PA environment.

Because of these three properties, RF PA design has often been described as a black art. Those endowed with these mystical powers of PA design methods rely upon different techniques of physical impedance synthesis to search for the optimum terminating conditions, while using indirect measurements to infer that an operating target or trade-off condition has been achieved. With a DC current meter and an RF power meter, one can infer optimality, without actually having to see the time-domain voltage and current signals of the RF PA. How these physical impedances are synthesized, and how the resultant measured data is used for matching network design, forms the essence of all RF PA design methods.

1.2 History of RF Power Amplifier Design Methods

Until the launch of practical commercial automated load-pull in the 1980s, two broad classes RF PA design methods were commonplace. The first class was based on a physical implementation of a variable impedance network. The input and output of the transistor were terminated by this variable impedance network, and while manually adjusting impedance, measurements showed performance that allowed identification of optimum impedance terminations to be located. The most common of these methods was the shunt-stub coaxial tuner, augmented by copper tape and an X-Acto knife for tuning in micro-strip media. Once optimum performance was identified, the transistor was removed and the input and output terminating impedances were measured with VNA, and then replicated by an appropriately designed matching network. Repeating this process several times facilitated construction of data contours superimposed on a Smith chart or rectangular impedance plane as an aid in identifying performance maxima and minima and the terminating impedances endowing the target performance.

The second class relied on analytical expressions, possibly supplemented with measured data or transistor data-sheet parameters, to approximate the optimum terminating impedance terminations. The most important of these methods is the Cripps method [1]. This method identifies an optimum load impedance, followed by design of a test-fixture whose impedances replicate those identified as optimum by the Cripps analysis.

Both classes require the impedance terminating the input and output of the transistor be known, at the fundamental frequency, and possibly the harmonic and baseband frequencies. The terminating impedances serves as an aid in contour construction, design of matching networks, or, often times, both. The key benefit of automated load-pull over these methods is *a priori* empirical determination of optimal impedance terminations and the ability to rapidly measure fully de-embedded performance. Automated load-pull resulted in a dramatic reduction in design cycle-time and post-fabrication tuning while simultaneously providing accessibility to RF PA design to a much larger group of engineers than before. In fact, the standard set by automated load-pull is so high that it has been difficult to displace it by newer methods, particularly EDA and time-domain methods, because the incremental improvement provided against the additional cost often does not lead to an appreciable difference in performance or reduction in design cycle-time.

1.2.1 Copper Tape and the X-Acto Knife

Exclusively for micro-strip media up to several GHz, use of an X-Acto knife and copper tape is a highly popular, legitimate, and effective method of impedance synthesis for RF PA design. A similar approach can be adopted in a MMIC environment using appropriate metal-mask design and a laser. Beginning with a double-sided substrate, a straight-edge is used to guide the X-Acto knife blade to remove cladding, synthesizing the top half of a micro-strip transmission line. Once the cuts are made, the cladding is peeled off, yielding a micro-strip line, usually 50 Ω. Full-wave electromagnetic (EM) simulation tools or closed-form approximations are used to determine the width of the top-metal conductor to establish its characteristic impedance. For EM simulations, Keysight Momentum is highly recommended, for its accuracy and ease-of-use, particularly how it handles the reference-plane for complex structures like that found in load-pull test-fixtures.

By placing one, or often two, shunt chip capacitors between the micro-strip line and ground-plane, the resulting multi-section low-pass matching network can transform down to a few ohms over bandwidth spanning most wireless standards. Depending on supply voltage, this method can be used to design an RF PA of 100 W or more. In addition, the X-Acto knife can be used to fabricate high impedance quarter-wave bias feeds to the transistor. Type N or 3.5 mm connectors are directly soldered the substrate, or, alternatively, the substrate can be reflowed or mechanically attached to a metal base to secure the connectors. The metal base provides necessary heat-sinking and mechanical strength.

1.2.2 The Shunt Stub Tuner

Because of its flexibility and ease of use, stub tuning is a popular method of RF PA design. It can be used to 10 GHz and well over 100 W. These advantages are offset by a narrow modulation bandwidth, high insertion loss, and unpredictable out-of-band effects. Stub tuners are generally available in single-stub, double-stub, and triple-stub formats. The single-stub tuner cannot match over the entire Smith chart, whereas the double-stub and triple-stub tuners can.

Although stub tuning is easier to use than using an X-Acto knife and copper tape, it is also less intuitive. When plotted on a Smith chart, the value of each shunt capacitor and their physical distance from the transistor leads reveals the impedance presented to the transistor. With a shunt stub, pre-calibrated graduation scribes must be placed on each stub or the stub is simply removed and measured on a VNA. By adjusting the radial distance from the center conductor of each stub, a transformation down to a few ohms can be made. To overcome matching and bandwidth limitations, the stub tuner is often cascaded with a pre-matching network on the test-fixture of the transistor, using either Type N or 3.5 mm connectors. The test-fixture also provides bias, DC blocking, and a heat-sink interface.

Closed-form expressions are available for stub tuners to determine what the radial positions must be to synthesize a prescribed impedance, though it is more convenient to simply move the radial stub position while monitoring RF and DC performance and record its impedance with a VNA. Contours can be fit or tuning can be carried out until optimum performance is achieved, upon which a matching network can be designed.

1.2.3 The Cripps Method

Of the analytical design methods, the Cripps method is the most intuitive, physical, and easiest to use. The Cripps method, published in 1983, is a large-signal empirical-analytical hybrid formulation that establishes the maximum power capability and gain of the transistor by considering its breakdown voltage and saturation current [1]. Parasitic effects, including drain-source capacitance, wire-bond inductance, and package lead delay, are removed, allowing contours to be created directly at the active die reference-plane.

Based on a combination of DC and RF data, usually from measurements and the transistor's data-sheet, respectively, the Cripps method provides remarkably good results. After extracting contours, and an associated large-signal load terminating impedance, an optimum input terminating impedance is established and matching networks are designed. This method yields similar results to measurement-based methods based, while having the advantage of directly providing contours and not requiring significant capital expense. As a sanity-check, the Cripps method often augments load-pull to establish consistency between s-parameters, large-signal characterization at 50 Ω, and the optimum Cripps load impedance.

1.3 The Load-Pull Method of RF Power Amplifier Design

The etymology of load-pull derives from the well-known tendency of an RF oscillator to shift its center frequency, or be pulled away from nominal, when its load impedance is altered, deliberately or otherwise. Hence, pulling the load impedance away from nominal has become associated with standard oscillator performance evaluation to assess frequency stability and spectral purity. In a similar fashion, the notion of deliberately and systematically varying the load impedance presented to an RF PA has become known colloquially as load-pull; it is both a noun and a verb and applies to the source as well as the load.

1.3.1 History of the Load-Pull Method

By presenting a variable source and load impedance to a transistor, the first published result of computer generated load-pull contours was by Cusack and Perlman [2]. Through a process of collecting data at various impedances, appropriately de-embedded, contours of power, gain, and efficiency were plotted using a contouring algorithm running on a main-frame computer, allowing optimum source and load impedances to be identified.

Between the publication of the Cusack paper in 1972, and the launch of practical commercial load-pull in the late 1980s, several individuals fabricated early versions of automated tuners, which included software and a computer interface. Stepper motors establish the magnitude and phase of the impedance by moving a probe inserted between the plates of slab-line. Both Secchi and Tsironis designed and fabricated early versions of automated tuners that formed the foundations of commercial automated load-pull at RCA and Focus Microwaves, respectively. For many years, Maury Microwave also sold manual slab-line tuners, which too ultimately formed the foundation for their commercial automated tuner product-line.

The first practical commercial load-pull was launched by the Sarnoff Laboratories of RCA in the mid-1980s, and by the early 1990s, load-pull was being deployed on a large scale throughout the world. The RCA architecture synthesized impedance using a dual-probe approach. Each of the two probes completely surrounded the center conductor and had only horizontal movement. Probe separation was adjusted to synthesize the reflection magnitude, and both probes were moved together (with the same separation) to adjust reflection phase.

Shortly following RCA, Maury Microwave and Focus Microwave both commercially launched automated load-pull tuners based on controlling the radial displacement of a single probe from the center conductor of a slab transmission line, coincident on a movable carriage to establish axial displacement from the device under test (DUT). Figure 1.1 illustrates the Focus Microwaves CCMT-50250,

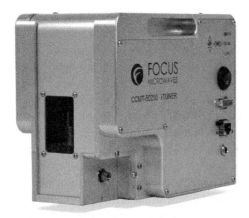

Figure 1.1 Contemporary microwave tuner spanning 2.5–50 GHz operating bandwidth. Source: Reproduced with permission of Focus Microwaves, Inc.

Figure 1.2 Contemporary microwave tuner spanning 600 MHz to 18.0 GHz operating bandwidth. Source: Reproduced with permission of Maury Microwave, Inc.

a contemporary state-of-the art passive tuner spanning 2.5–50 GHz operating bandwidth that could be used, for example, for on-wafer load-pull. Figure 1.2 illustrates the Maury Microwave XT982GL01, a contemporary state-of-the art passive tuner spanning 600 MHz to 18.0 GHz, commonly used for high-power fixture-based load-pull.

Though these early architectures were each based on electromechanical tuners, their data acquisition methods were radically different. The Sarnoff system used a search algorithm to locate impedance states displaying target performance, such as a contour 2 dB below maximum gain. In contrast, Focus and Maury performed load-pull over a range of *a priori* user-selected points and fitted contours *a posteriori* to the measured data in PC-based graphics package. This evidently benign change, in fact, represented an extraordinary shift in load-pull philosophy as it resulted in an order of magnitude reduction in data acquisition time.

The initial systems from these companies were based on a passive architecture, meaning a transmission line used with motor-controlled probes and carriages that dynamically modified the characteristic impedance of the transmission line to present an arbitrary impedance. At the time of the writing of the present book, passive load-pull is the dominant architecture, though active load-pull is a well-established compliment, particularly for on-wafer and high-speed applications.

Takayama was the first to propose an alternative to the passive load-pull architecture, using an amplifier and a phase shifter to synthesize a virtual impedance at the load fundamental [3]. Active load-pull has the primary advantage of overcoming probe losses in on-wafer applications, particularly at mm-wave frequencies, where probe loss is substantial. Since Takayama, and in particular since development of real-time high dynamic-range time-domain data acquisition technology in the 2000s, active load-pull has become commercially practical.

1.3.2 RF Power Amplifier Design with the Load-Pull Method

The load-pull method systematically varies the source and load impedance at the input and output of the transistor, allowing contours to be superimposed over a Smith chart that identify optimum impedances subject to some criteria. Figure 1.3 illustrates a generic load-pull system with input and output terminations defined as Γ_{Source} and Γ_{Load}. By replicating the optimum impedance terminations at all relevant mixing products, the performance exhibited by a transistor in load-pull will match that exhibited in the final design, net of matching network insertion loss. In contrast to methods that approximate performance or rely on assumptions, the power of load-pull follows from identification of optimum impedance terminations under realistic stimulus and a well-defined impedance reference-plane at both the input and output of DUT. It is the only empirical method that provides the designer with maximum performance and trade-off conditions.

To design with load-pull, a structured, often iterative, process is followed to reveal optimum performance and the associated terminations. The process begins by identifying the key parameters that are to be optimized. Generally these are the parameters whose contours will need to be plotted, in the source impedance plane,

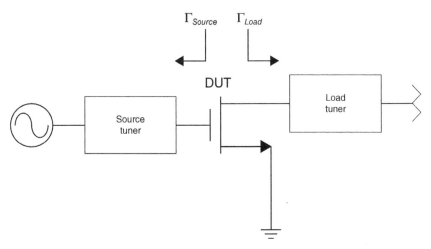

Figure 1.3 Block diagram of a generic load-pull system illustrating key impedance definitions.

load impedance plane, or both. Often times multiple criteria will exist that must be traded off against each other or against a specific matching network design. For example, in the load impedance plane, output power, gain, and power-added efficiency (PAE) must be traded off against each other to establish absolute performance and the associated terminating impedance.

It is commonly, and incorrectly, thought that Γ_{Source} and Γ_{Load} presented by the terminating networks shown in Figure 1.3 are necessarily complex conjugates of Γ_{In} and Γ_{Out}, respectively. Not only is this incorrect, based on the previous definition of the RF PA, but superposition does not hold under large-signal conditions. Instead, the terminating impedances presented are those load-pull conditions that allow the prescribed performance to obtain, while naturally interacting with the transistor input and output impedance over all relevant mixing products. Only under small-signal conditions is the notion of maximum power transfer and conjugate termination meaningful.

1.4 Historical Limitations of the Load-Pull Method

First-generation automated load-pull offered a significant advance over stub-tuning and the X-Acto™ knife, though many limitations nevertheless existed in initial commercial releases, particularly the impedance range necessary for wireless applications and harmonic tuning for high-efficiency modes such as Class F. With the launch of second-generation pre-matching and harmonic load-pull technology, along with extensive growth in test-fixture sophistication and understanding, virtually all of these limitations have been resolved, ensuring that load-pull can emulate all but the most demanding environments.

At its inception, load-pull faced the following limitations listed subsequently. Each of these has either been fully resolved or have identified work-around that mitigate their impact on the usefulness of load-pull. In most of these limitations, a substantial body of empirical and analytical literature has evolved, rigorously treating limitations, by providing solutions or guidance in avoiding impairment or difficulty. This section addresses each of these limitations in sequence to demonstrate that load-pull is a reliable method for systematic, accurate, and fast RF PA design built on a rigorous analytical foundation:

- Impedance range
- Harmonic tuning
- Power
- Bandwidth
- Linearity impairment
- Rigorous error analysis
- Acoustically induced vibrations

1.4.1 Minimum Impedance Range

First-generation commercial load-pull presented a minimum impedance of 5 Ω at the common wireless bands at 800 and 1900 MHz, which was not low enough for GSM and CDMA handset impedances of 2 and 4 Ω, respectively. Base-station Si BJT and laterally diffused metal oxide semiconductor (LDMOS) transistors in the 50 W range, popular at the time, required an impedance of 2 Ω, with sub-1 Ω impedance required for the +200 W LDMOS technology available presently. Optimum linearity tuning also requires sub-1 Ω source impedance, depending on the nature of the pre-matching embedded within the package. To work around this first-generation limitation, scaling was used, with limited success, because of distributed effects over the transverse axis of the semiconductor die, wire-bonding, and package delay.

As the first practical solution to the low impedance problem, pre-matching was rigorously applied in 1997, using two-tier TRL de-embedding to synthesize 0.2 Ω at 800 and 1900 MHz in micro-strip media [4]. This approach relies on the distributed impedance transformer, starting with a single-section quarter-wave line and increasing in sophistication to include multi-section Chebyshev and Hecken tapers [5]. This method has several advantages, including significant impedance transformation to the sub-1 Ω region, enhanced bandwidth, power capability enhancement of the load-pull system, and a test-fixture environment that is many cases identical to the final design, thereby reducing uncertainty. A limitation of pre-matching, particularly for single-section quarter-wave trans-formers, is bandwidth, requiring different fixtures to cover each of the common wireless bands.

Second-generation load-pull technology offered a major advance in low impedance load-pull with dynamic pre-matching. With multi-probe technology, one probe in a mechanical tuner can act as a dynamic pre-match to a second probe, thereby allowing a frequency agile impedance synthesis system that rivals multi-section distributed pre-matching in minimum impedance. For all but the most demanding low impedance applications, multi-probe technology is the preferred approach for most low impedance applications, though it is often paired with pre-matching for additional impedance transformation and form-factor matching to the package lead-width.

1.4.2 Independent Harmonic Tuning

High-efficiency operating classes, such as Class F, require high-reflectivity har-monic terminations for proper operation. Harmonic terminations also influence linearity. Because initial commercial load-pull technology could not indepen-dently tune the fundamental and harmonic impedances, these systems were incapable of replicating the conditions for high-efficiency operation, thereby restricting operation to standard operating classes or resulting in linearity error between load-pull and final design due to harmonic termination differences.

The first harmonic load-pull system was created by inserting a duplexer in front of a mechanical tuner, with the coupled port presented as the load to the DUT, the low frequency port terminated by fundamental tuner, and the high frequency port terminated by a coaxial sliding short transmission line. With the duplexer providing isolation between the fundamental tuner and the second harmonic slid-ing short, the second harmonic could be optimized, though insertion loss lim-ited the practicality of this method. Nevertheless, this approach was deployed as first-generation load-pull technology.

A similar approach, with higher second harmonic matching range, was to insert a half-wave resonator in the mechanical tuner, in front of the fundamental probe. While offering nominally higher reflection at the second harmonic, this method suffered from an extremely narrow bandwidth and fundamental coupling.

Expanding on the multi-probe technology for low impedance synthesis, second-generation load-pull implements harmonic tuning to fourth harmonic by mathematically synthesizing uncoupled impedances from multiple probes in one mechanical tuner. This method is a substantial improvement over previous methods, allowing not only harmonic tuning, but independent impedance synthesis at up to four arbitrary frequencies, and is now the standard approach for passive harmonic load-pull.

1.4.3 Peak and RMS Power Capability

Electric field breakdown and localized Joule heating are the two physical mechanisms that establish an upper bound on peak power and average power for load-pull, respectively.

For an automated coaxial tuner, the electric field between the shunt probe and center conductor creates capacitive coupling inducing susceptance, subsequently seen as a change in the magnitude of the reflection coefficient of the tuner. Probe displacement toward the center conductor increases the electric field intensity, which breakdowns at 31 kV/cm in vacuum. While not necessarily catastrophic to the tuner, breakdown oxidizes the probe at the source of the arc, changing its physical constitution and permanently altering its impedance transforming properties. Probe displacement also induces circulating current to ground, the physical basis of impedance transformation, creating heat due to metal loss within the tuner assembly. Large current distributed over a small metal volume, sustained over time intervals of milliseconds, leads to a significant temperature increase, inducing physical expansion of tuner components. While not causing permanent damage, this metal expansion can cause an uncalibrated change in the tuner impedance.

Current-generation load-pull technology is capable of multi-hundred watt load-pull by judicious choice of pre-matching fixture design coupled with multi-probe tuner technology that reduces peak electric field and heating by allocating matching over several sections. Proper choice of the RF connector is also necessary, as poor connector quality or a damaged connector each represent substantial insertion loss, particularly if a current maxima node exists within the connector, near a discontinuity or support bead.

1.4.4 Operating and Modulation Bandwidth

The load-pull tuner exhibits an intrinsic frequency response that establishes its operating bandwidth. The operating bandwidth and the instantaneous, or modulation, bandwidth, are the two most important types of bandwidth relevant for the load-pull tuner since they establish the usable tuner bandwidth and maximum modulation bandwidth of the signal applied to the tuner.

The operating bandwidth of a coaxial tuner is limited by the minimum *VSWR* that can be generated over an associated frequency range. Using the single-probe structure with a single matching element, it is possible to achieve a multi-octave operating bandwidth. Periodic, or corrugated, probe technology can deliver wider operating bandwidth or, for an equivalent single-probe frequency response, higher *VSWR*. Second-generation tuner technology consists of two distinct probes that deliver over a decade of operating bandwidth. The operating bandwidth of a wave guide tuner is usually limited to the frequency range of its dominant mode, usually the TE_{10} mode.

Matching networks used to extend the minimum impedance range are also used to extend bandwidth, particularly for harmonic load-pull. Multi-section quarter-wave lines or tapered lines provide multi-octave bandwidth to enable not only low impedance for fundamental load-pull but also the high *VSWR* necessary for harmonic load-pull.

1.4.5 Linearity Impairment

A linear network corrupts a signal when its complex spectral components are exposed to nonconstant group delay within the bandwidth occupied by the instantaneous frequency deviation of the modulation. The most common form of corruption is spectral asymmetry about the carrier, thereby impairing important out-of-band signal quality metrics such as third-order intermodulation (IM_3) ratio and adjacent channel power ratio (ACPR). Linear distortion is often confused with asymmetry caused by nonlinearities of the PA, hence its importance in load-pull. Nonconstant group delay is usually caused by parallel resonance in the bias network, often between deliberate insertion of capacitance as an RF short termination and a colocated large electrolytic capacitor operating above its first resonance, and thus appearing as an inductor. In wideband modulation applications, greater than approximately 150 MHz, bias network inductance interacts with internal pre-matching MOS capacitance of the DUT, creating parallel resonance. Other forms of memory, with dominant time constants near the modulation rate, will also impair signal quality. These include electrothermal memory and material-related memory, such as bulk and surface states in gallium arsenide (GaAs) and gallium nitride (GaN).

Video bandwidth (VBW), a terminology holdover from the television era, is the name commonly applied as the definition of the bandwidth spanning DC to the first parallel resonance of the bias feed, upon which group delay becomes sufficiently nonlinear to impair signal quality. Impairment of IM and ACPR as an artifact of the load-pull process reduce the apparent PAE and average power capability of the RF PA, and hence must be minimized for load-pull to be effective as a design method.

First-generation load-pull technology suffered more from an inadequate understanding of signal quality impairment caused by nonlinear group delay rather than physical tuner performance. Indeed, many RF PA engineers express disbelief when first confronted with the notion that the bias network, which is linear and passive, can impair a signal in a fashion identical to a nonlinear RF PA. By the time second-generation load-pull technology was released, a substantial body of knowledge had been developed, based on linear-phase network synthesis theory, Volterra series, and an *in situ* VBW measurement method from Noori [6]. By properly allocating impedance transformation between the tuner and a pre-matching network, coupled with careful management of inductance external to the DUT, such as that exhibited by power supply leads, a VBW in excess of 300 MHz is possible. This is more than adequate for 4G and 5G wireless standards using digital pre-distortion (DPD).

1.4.6 Rigorous Error Analysis

With the exception of tuner repeatability, a rigorous error analysis method for load-pull did not exist for nearly 10 years after its commercial launch in 1988, being a major contributor to its initial skepticism. A rigorous error analysis considers both systematic and random contributions. The systematic error term is time-invariant, and once identified and understood, can be effectively removed from influencing the total error performance of the load-pull system. The influence of the random error term cannot be entirely removed from its effect on accuracy, but once its physical origins are understood, it can be controlled to an arbitrarily negligible, or at least acceptable, level. More importantly, a simple test can be performed, much like with a VNA calibration verification, to quantify the random error term and relate it to measurement error of parameters such as power, gain, and PAE.

Two methods are available for error analysis. The most important of the two, commonly called the ΔG_t method, was invented by Sevic and Burger [7]. By comparing the transducer gain a load-pull system against the transducer gain using s-parameters, the method provides an upper bound on gain error. This error can further be related to power and PAE error, thereby providing the user with a quantitative assessment of not only uncertainty for gain but its effect power and PAE uncertainty. The utility of this method is evident when noting that tuner repeatability cannot provide a quantitative prediction of parametric uncertainty. By performing the ΔG_t method over power and impedance, error surfaces can be generated to fully understand the influence of random error on load-pull.

1.4.7 Acoustically Induced Vibrations

The mechanical inertia of the components internal to a passive-mechanical tuner may induce acoustically coupled vibrations to the wafer-probe used to make electrical contact to the wafer or other nanoscale device. This is particularly an issue with silicon processes using aluminum, such as CMOS flows, due to oxidation and its resultant effect on contact resistance.

Using an RF cable to connect the tuner to the probe probes substantial isolation although it results in *VSWR* contraction due to probe insertion loss. Second-generation load-pull technology, using multi-purpose technology, overcomes the acoustically coupled problem by eliminating movement entirely of the probe carriage, which has most of the inertia, and instead moving only three probes. This method completely eliminates contact problems.

1.5   Closing Remarks

The RF PA is distinguished by operation in the large-signal regime, the presence of distributed effects, and a substantial temperature rise above ambient. Operation in the large-signal regime forbids assumption of the superposition principle, so that *a priori* assessment of DUT performance cannot be extrapolated from linear two-port characterization, as assumed in the weakly nonlinear regime and *s*-parameters. Optimum operation of the DUT relies on identification of the optimum impedance terminations not only at the fundamental but all relevant mixing products as well. Classical RF PA design methods have therefore heavily relied on empirical methods often involving proxy measurements to infer the actual operating state of the RF PA. The automated load-pull method of RF PA design enables optimal transistor terminating impedances to be systematically, accurately, and rapidly identified without resorting to slow and uncertain trial and error associated with historical procedures.

References

1 Cripps, S. (1983). A novel method of predicting FET power contours. IEEE International Microwave Symposium.

2 Cusack, J. and Perlman, B. (1972). A method of plotting load-pull contours using a computer. IEEE International Microwave Symposium.

3 Takayama, I. (1976). A new method of active load-pull. IEEE International Microwave Symposium.

4 Sevic, J.F. (1996). A rigorous two-tier fixture calibration method for low impedance load-pull. Automatic RF Techniques Group Conference.

5 Hecken, R.P. (1972). A near-optimum matching section without discontinuities. IEEE Transactions on Microwave Theory and Techniques.

6 Noori, B. (2004). A simple method of calculating VBW in-fixture. IEEE Radio and Wireless Symposium.

7 Sevic, J.F. and Burger, K. (1996). Rigorous error analysis for low impedance load-pull, from Internal Notes at Qualcomm.

2

Automated Impedance Synthesis

The method of impedance synthesis employed by a load-pull system plays a central role in establishing its overall performance, particularly the impedance range, operating bandwidth, power capability, harmonic tuning capability, and measurement speed. While no one method, or architecture, excels in all dimensions, the passive-mechanical comes closest in terms of simultaneously meeting most performance goals while retaining a great deal of flexibility at reasonable cost. Electromechanical impedance synthesis is currently the most common impedance synthesis method, and is the focus of the present chapter, though the methods treated can be applied as well to load-pull with alternative impedance synthesis methods, such as active and hybrid.

Active impedance synthesis is based on either the active-loop or active-injection architectures. Active load-pull offers unity reflection coefficient magnitude capability, or larger, that compensates for loss typically encountered by wafer-probes and cables. It is ideally suited for on-wafer mm-wave applications, where probe losses cannot be neutralized by with passive impedance synthesis, and for ultra-fast load-pull, where impedance can be synthesized rapidly using controlled sources. Though not necessary, active load-pull is often coupled with time-domain signal acquisition hardware, for harmonic load-pull, forming the hybrid architecture.

The present chapter provides an introductory treatment of impedance synthesis methods for load-pull using including electromechanical passive impedance synthesis and active impedance synthesis, with emphasis placed on internal operation and associated benefits and limitations of each method. Common tuner specifications are introduced, along with associated manifestations of performance impairment. The chapter closes with discussion of advanced methods harmonic impedance synthesis and sub-1 Ω load-pull.

The Load-Pull Method of RF and Microwave Power Amplifier Design, First Edition. John F. Sevic.
© 2020 John Wiley & Sons, Inc. Published 2020 by John Wiley & Sons, Inc.

2.1 Methods of Automated Impedance Synthesis

2.1.1 Passive Electromechanical Impedance Synthesis

Passive electromechanical impedance synthesis is based on the insertion of a metallic probe in the transverse plane of a wave-guiding structure, as illustrated in Figure 2.1, terminated at each end by either a coaxial or rectangular waveguide interface. For operation to 50 GHz, the wave-guiding structure is based on either a slotted coaxial transmission line or parallel-plate transmission line operating in *TEM* mode; for higher frequency operation, the transmission line is based on a slotted rectangular waveguide operating in the TE_{10} mode. Coaxial tuners are terminated by industry-standard coaxial transitions, including Type N, 3.5 mm, 2.4 mm, and 7–16,[1] with rectangular waveguide tuners terminated by industry-standard WR-*xy* waveguide transitions.

The probe behaves as shunt stub of variable susceptance, depending on the ratio of its displacement from the center conductor, d, to the operating frequency. It appears either inductive or capacitive, except at series resonance, where it emulates a near-lossless short, or at parallel resonance, where it emulates an open. Figure 2.2 illustrates a common probe used in the passive electromechanical tuner. The edges of the probe where it establishes initial proximity with the center conductor are tapered to reduce the rate of change of the reflection coefficient magnitude with probe displacement. To first-order, the magnitude of the reflection coefficient generated by the probe is inversely proportional to probe displacement d, graphically illustrated by the Smith chart of Figure 2.3. The waveguide tuner uses instead an E-plane vane that rotates to produce a reflection.

Probe dimensions are optimized for a specific frequency range, and can either contact the waveguide wall or remain a small distance away. The former is commonly called a contacting probe, whereas the latter is referred to as a

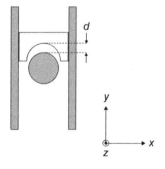

Figure 2.1 Transverse section view of electromechanical tuner slab-line illustrating probe placement and its displacement for synthesis of an arbitrary impedance. To first-order, probe displacement from the center conductor (along the *y*-axis) represents the magnitude of the reflection coefficient and its longitudinal displacement (along the *z*-axis) from an arbitrary reference-plane, usually the physical end of the tuner nearest the DUT, represents phase.

---

1 The 7–16 connector is a high-power RF connector composed of an inner conductor of 7 mm diameter and outer conductor of 16 mm diameter.

Figure 2.2 Typical probe used in the passive electromechanical tuner. Each square-grid represents 1 cm. Source: Reproduced with permission of Focus Microwaves, Inc.

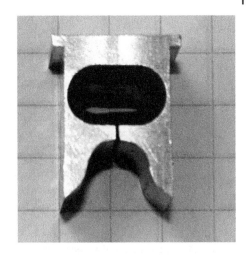

Figure 2.3 Reflection coefficient vector seen at tuner calibration reference-plane illustrating relationship between probe displacement and carriage displacement and associated magnitude and phase.

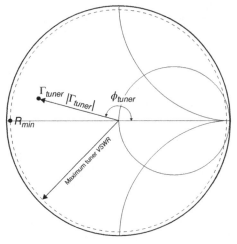

noncontacting probe. Contacting probes are usually machined from brass to provide a low-friction interface to the transmission line wall, with noncontacting probes made of anodized aluminum. As a general rule, the contacting probe has slightly higher bandwidth than the noncontacting probe, and will generally not exhibit resonances under maximum *VSWR* generation. In contrast, the noncontacting can exhibit improved repeatability and better long-term reliability.

Figure 2.4 shows an internal view of a modern multi-probe coaxial tuner capable of fundamental and harmonic impedance synthesis. To control the phase of the generated reflection coefficient, the probe is mounted on a carriage that moves in parallel to the longitudinal axis of the transmission line. To first-order, the phase of the reflection coefficient is linearly proportional to its displacement from an

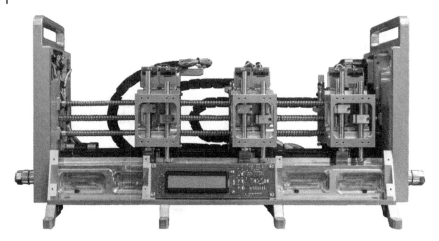

Figure 2.4 Internal view of a modern electromechanical tuner with three carriages. A multi-probe tuner of this style is capable of independent fundamental and harmonic impedance synthesis, as well as frequency-agile dynamic pre-matching. Source: Reproduced with permission of Focus Microwaves, Inc.

arbitrary calibration reference-plane, usually at the side of the tuner in proximity to the transistor, to reduce insertion loss. Figure 2.3 illustrates the definition of phase of the reflection coefficient produced by the tuner, looking back from the DUT reference-plane.

The frequency at which the transmission line is electrically $\lambda/2$ establishes its lower operating frequency; the upper frequency limit is related to probe size. Limit switches prevent carriage displacement into either end of the tuner, in addition to probe displacement creating physical contact with the center conductor, eliminating the possibility of a DC short and serious physical damage.

Skin effect losses at the probe and transmission line surfaces place an upper limit on the maximum *VSWR* the tuner can generate, typically 200 : 1 at wireless frequencies and 25 : 1 at mm-wave frequencies. Adding to the upper limit on reflection coefficient is an associated rapid increase in the available loss of the tuner at maximum *VSWR*. An increase in available loss requires a corresponding proportional increase in the driver power amplifier (PA) of the load-pull system, to drive the DUT. In extreme cases, tuner self-heating can alter its reflection coefficient due to center conductor sag; an indirect benefit of pre-matching is partitioning of impedance transformation over multiple sections, leading to lower circulating current, and lower insertion loss, and higher net *VSWR* at the DUT reference-plane.

2.1.2   The Active-Loop Method of Impedance Synthesis

The active-loop method of impedance synthesis uses one of two fundamental architectures to launch toward the DUT an amplitude and time-delayed replica of its output wave, thereby synthesizing a virtual load impedance based on wave interference phenomena. Because the amplitude of the reflected wave is arbitrary, within the peak envelope power (PEP) capability of an external source, it can be scaled to exhibit arbitrary power so that losses between the synthesis reference-plane and transistor reference-plane are effectively neutralized. This unique property of active impedance synthesis enables generation of a reflection coefficient greater than unity magnitude, and is therefore capable of placing a reflection coefficient at the DUT reference-plane anywhere on the interior or boundary of the Smith chart, as well its exterior. Loss neutralization makes active impedance synthesis ideal for on-wafer mm-wave applications, where probe and cable losses are substantial.

The *VSWR* range and frequency range of active impedance synthesis is established primarily by the performance of the associated hardware, particularly the driver PA, which in fact can be much larger in PEP capability than the required DUT PEP. This property a result of the potentially large mismatch between the 50 Ω source and the virtual impedance seen by DUT, which can be 1 Ω or less. This is illustrated by Figure 2.5, showing the ratio of driver PA PEP to DUT PEP versus minimum impedance synthesized at the DUT reference-plane. For a target impedance of 1 Ω, an associated power headroom of approximately 12 dB is required, depending on loop coupling efficiency. The active-loop method is easily extended to harmonic load-pull by adding additional harmonic synthesis loops.

Figure 2.5   Fundamental reference PA power capability normalized by expected DUT PEP capability versus required synthetic load impedance at DUT reference-plane for active-loop. The four trajectories illustrate 0–3 dB insertion loss, in 1 dB steps.

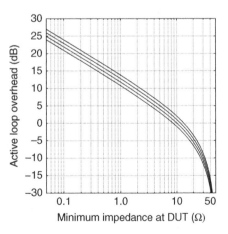

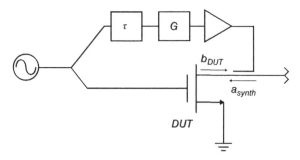

Figure 2.6 The fundamental feed-forward active-loop impedance synthesis architecture, due to Takayama, and often referred to as the split-signal method [1].

The first published active-loop load-pull system was due to Takayama [1]. Figure 2.6 illustrates the fundamental feed-forward Takayama architecture, also referred to as the split-signal method, since the reference signal for impedance synthesis is acquired before at the source-side of the DUT versus the load-side. The driver PA provides gain G, and the delay block provides frequency-independent time delay τ.

The reflection coefficient presented to the load of the *DUT* of Figure 2.6 is defined as the ratio

$$\Gamma_{load} = \frac{a_{synth}}{b_{DUT}(\Gamma_{source}, P_{avs})} \tag{2.1}$$

where synthesized traveling wave a_{synth} is an amplitude and time-delayed replica of traveling wave $b_{DUT}(\Gamma_{source}, P_{avs})$ launched by the *DUT* toward the virtual load, and is shown to be a function of available source power and source impedance on the input side of the *DUT*. Since the magnitude and phase of Γ_{load} are established by both a_{synth} and b_{DUT}, dynamic AM-AM and AM-PM of the *DUT* influence the apparent reflection coefficient. For this reason, the feed-forward active-loop architecture requires a mechanism to compensate for instantaneous changes in AM–AM and AM–PM, usually implemented as a slowly varying power control feedback loop. The feed-forward active-loop architecture offers unconditional synthesis stability, though practical power control loops have potential for low frequency oscillations, within the loop bandwidth of the power control transfer function.

To eliminate the dynamic AM–AM and AM–PM compensation problem, consider the fundamental feed-back active-loop architecture of Figure 2.7. By sampling the launched wave of the transistor after amplification, versus before, the feed-back architecture eliminates the need for dynamic AM–AM and AM–PM compensation, but because of the negative feedback, the potential for oscillation exists where the unity loop gain phase shift is 360°.

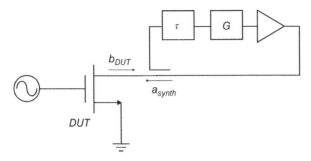

Figure 2.7 The fundamental feed-back active-loop impedance synthesis architecture.

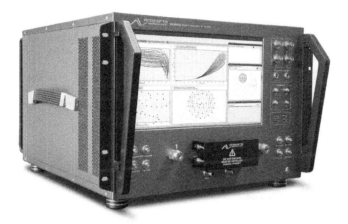

Figure 2.8 Modern active-loop load-pull system capable of simultaneous fundamental, harmonic, and baseband characterization. Source: Reproduced with permission of Maury Microwave, Inc.

Source-pull and characterization of wide-band modulation with active-loop method of impedance synthesis can present challenges that in general do not exist with passive electromechanical synthesis. Source-pull with active-loop requires computationally-intensive iteration, and is often substantially slower than passive electromechanical source-pull. Wide-band modulation stimulus with active load-pull requires high-speed ADC hardware to support data rates associated with 4G and 5G, and can provide de-embedded time-domain data as well. Figure 2.8 illustrates an active-loop load-pull system, based on wideband ADC technology, capable of supporting 1 GHz modulation bandwidth commonly associated with base-station digital pre-distortion (DPD) applications.[2]

_____

2 The reader is referred to Maury Application Notes 5A-046 and 5A-044 for further details on active-loop load-pull.

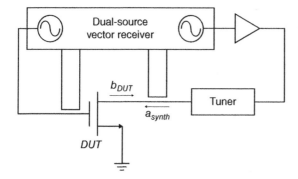

Figure 2.9 The fundamental active-injection impedance synthesis architecture.

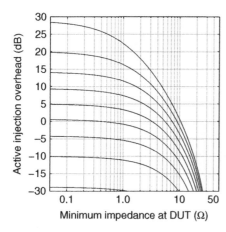

Figure 2.10 Fundamental reference PA power capability normalized by expected DUT PEP capability versus required synthetic load impedance at DUT reference-plane for active injection. The nine trajectories illustrate passive pre-matching from 0.1 to 0.9, in 0.1 steps.

2.1.3 The Active-Injection Method of Impedance Synthesis

The active-injection method of impedance synthesis retains the benefits of the active-loop method, including an arbitrary reflection coefficient and high-speed data acquisition, while reducing reliance on the high-power loop PA required in both the feed-forward and feed-back architectures of Figures 2.6 and 2.7, respectively. By inserting an electromechanical tuner, as shown in Figure 2.9, the active-injection method substantially reduces loop PA PEP requirements for fundamental load-pull and eliminates the need for harmonic loop control and additional sources. Figure 2.10 illustrates pre-matching tuner influence on PEP requirements. For example, to generate 1 Ω with the active-injection method, with tuner pre-match set for 0.4, approximately 2 dB of PEP headroom is required, an order of magnitude less than the active-loop method.

The vector receiver of Figure 2.9 measures in real-time the load reflection coefficient synthesized by the combination of the pre-matching tuner and amplitude-

and phase-controlled source internal to the vector receiver. The load reflection coefficient is expressed as

$$\Gamma_{load} = \frac{a_{synth}(\Gamma_{tuner}, \kappa_{source}, \phi_{source})}{b_{DUT}} \qquad (2.2)$$

where Γ_{tuner} is the pre-matching reflection coefficient generated by the tuner, κ_{source} is the relative amplitude of the vector receiver source, and ϕ_{source} is its phase relative, ultimately, to the phase of b_{DUT}. Each of the three parameters in the numerator of Eq. (2.2) are controlled via an external application.

The operating bandwidth of both active-loop and active-injection is established largely by the bandwidth of the loop PA, and for active-injection the frequency range of the vector receiver. Because high-power PAs tend to have reduced bandwidth, replicating the operating bandwidth of the passive electromechanical method will require a banded approach, while multiple banded PAs will require a new calibration each time a different PA is used. Since active-injection substantially reduces loop PA PEP requirements, active-injection will offer an improvement in dynamic frequency agility and operating bandwidth over both architectures comprising the active-loop method.

Figure 2.11 shows the same independent variable as Figure 2.10, plotted instead against the magnitude of the reflection coefficient at the DUT reference-plane, for various tuner pre-matching coefficients, with an assumed injection loss of

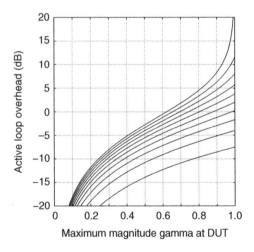

Figure 2.11 Fundamental reference PA power capability normalized by expected DUT PEP capability versus required synthetic reflection coefficient magnitude at DUT reference-plane for active-injection impedance synthesis. Insertion loss between the active-injection reference-plane and the DUT reference-plane is assumed to be 2.0 dB, with the initial trajectory set for 0 magnitude pre-match and each subsequent trajectory an increase of 0.1 up to 0.9. The reflection coefficient is normalized to 50 Ω.

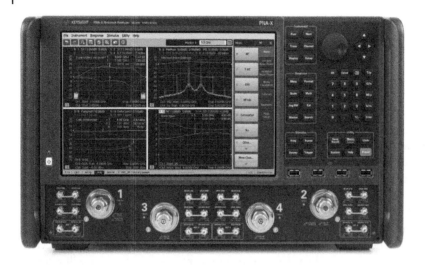

Figure 2.12 Modern high-performance vector network analyzer with four ports and time-domain capability. Source: Reproduced with permission of Keysight, Inc.

2.0 dB. Two trends follow from this illustration. As the magnitude of the reflection coefficient approaches unity, absence of pre-matching shows the required power increases rapidly to impractically large power requirements for common high-power wireless applications. Similarly, however, for modest pre-matching, the loop PA power required is practical. In a typical active-injection application, an electromechanical tuner is paired with a multichannel vector receiver, for example the Keysight PNA-X illustrated in Figure 2.12.

2.2 Understanding Electromechanical Tuner Performance

This section treats performance specification of electromechanical tuner performance, especially those relevant for the load-pull method of RF PA design, and their relation to physical design philosophy and constraints. For alternative applications with an electromechanical tuner, such as noise-parameter extraction, the relevant importance rank will be slightly different.

2.2.1 Impedance Synthesis Range

Impedance synthesis range, typically interpreted to mean the maximum *VSWR* the tuner can generate, is a key performance specification as it determines the range of possible impedances the tuner can generate. Load-pull of an RF transistor designed for power applications will generally require an impedance on the low impedance quadrant of the Smith chart, so that the minimum impedance the tuner generates is useful in establishing its ability to capture the minimum required load-line. Taking the minimum impedance to be real-valued, the minimum impedance the tuner will generate is related to its *VSWR* by

$$R_{min} = \frac{50\ \Omega}{VSWR} \tag{2.3}$$

where *VSWR* is the tuner *VSWR* at a given frequency and R_{min} is the minimum impedance associated to the *VSWR* at that frequency, as defined in Figure 2.3. At the common wireless band spanning 400 MHz to 6 GHz, R_{min} will range from 3 to 5 Ω, illustrating the need for pre-matching for some applications.

Maximum *VSWR* is limited by the proximity of the probe to the center conductor, illustrated in Figure 2.1, and tuner loss. Since the probe behaves as a shorted shunt stub, its susceptance versus displacement *d* can be approximated by

$$B_{short} = -jY_o \cot \beta z \tag{2.4}$$

where Y_o is the characteristic susceptance of the probe, β is the propagation constant, and z is the displacement from the center conductor surface. To assess the change in susceptance of the probe with respect to displacement, the derivative of Eq. (2.4) can be expanded around the first term of its Taylor series, for small βz, to yield

$$\frac{\partial B_{short}}{\partial z} \approx \frac{jY_o}{\beta z^2} \tag{2.5}$$

showing that when the probe is in proximity of the center conductor surface, its susceptance changes quadratically with displacement. Because of this behavior, the magnitude of the reflection coefficient rapidly changes as the probe approaches the center conductor surface, and generally the repeatability will degrade. As a general rule, it is advisable to rigorously validate the maximum *VSWR* range of the tuner – usually referred to as the outer-ring, since it is the circle representing maximum *VSWR* – to ensure repeatability requirements are met. Often times the outer-ring, as defined in Figure 2.3, is simply avoided and pre-matching is used.

An additional consequence of a quadratic susceptance function is a corresponding quadratic increase in circulating current in proximity of the center conductor surface. The stub behaves essentially as a shunt capacitor, and as its value increases to increase the magnitude of the reflection coefficient, surface metal loss of the probe, center conductor, and waveguide wall become appreciable.

2.2.2 Operating Bandwidth

Operating bandwidth is a key performance specification of the passive electromechanical tuner as it establishes minimum and maximum frequencies over which a minimum tuner *VSWR* is specified. The operating bandwidth will in turn determine the number of tuners necessary to provide a sufficient frequency span for a particular application, though it is often the case a single tuner provide adequate bandwidth for many applications. Harmonic load-pull is also influenced by the operating bandwidth, as the maximum operating frequency places an upper limit the harmonic order for load-pull, and consequently places also an upper bound on the maximum fundamental operating frequency.

The modern electromechanical tuner typically uses two distinct probes, sharing a common carriage, to provide a synthetic operating bandwidth exceeding a decade. The transition frequency from the low-frequency probe to the high-frequency probe is defined commonly as the crossover frequency, and is generally transparent to the user. Figure 2.13 illustrates a typical *VSWR* frequency response trajectory, illustrating the minimum operating frequency, the maximum operating frequency, and cross-over frequency, with 20 : 1 chosen commonly as the minimum associated *VSWR* to define the operating bandwidth. The peak in-band *VSWR* is generally substantially larger than band-edge or crossover *VSWR*, with *VSWR* and operating bandwidth usually traded off one another.

The practical physical operating bandwidth of the slab-line tuner is limited by its length, for the lowest frequency, particularly the ability to maintain slab-line and center conductor uniformity in their transverse plane. Stepper-motor resolution, and its associated influence on minimum phase resolution, and non-*TEM* propagating modes, establish limits on the highest practical operating frequency. As a practical boundary, 100 MHz is the lower frequency limit, while 50 GHz remains the upper frequency limit, established by the 2.4 mm coaxial transmission line. Adding stepper motor resolution as a constraint, with physical length, places the current maximum operating bandwidth for a single

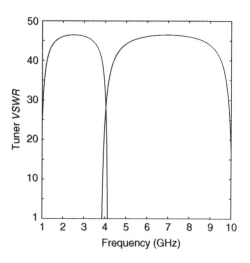

Figure 2.13 Typical dual-carriage tuner *VSWR* response versus frequency, illustrating operating bandwidth and crossover frequency, in this example being 4 GHz. The operating bandwidth is limited to approximately 9 GHz in this example, illustrated by the 15 : 1 *VSWR* at 10 GHz.

tuner from approximately 100 MHz to 2 GHz and 2–50 GHz. The operating bandwidth of the waveguide tuner is established by its dominant mode, usually the TE_{10} mode. Lumped-element tuners have been manufactured for VHF and HF applications and have been applied in a laboratory environment for CDMA baseband termination optimization [2, 3].

2.2.3 Modulation Bandwidth

Modulation bandwidth, or equivalently, video bandwidth (VBW), specifies the maximum instantaneous bandwidth over which group delay is sufficiently constant to maintain time alignment of the spectral components of the modulation applied to the DUT. Modulation bandwidth is a critical performance specification as signal impairment due to nonconstant group delay is manifested as spectral asymmetry in IM or adjacent channel power ratio (ACPR). Maintenance of load-pull system transparency is vital with the wideband modulation typically associated with WCMDA, 4G, 5G, and multi-carrier power amplifier (MCPA) applications.

The physical origin on the upper limit of modulation bandwidth is due to electrical memory induced by the process of impedance transformation. Memory phenomena in load-pull is particularly nefarious as it masquerades as spectral regrowth due to transistor nonlinearity, yet is generally indistinguishable from, and cannot be separated from, transistor nonlinearity. Maximization of the modulation bandwidth is therefore essential to enhance the transparency of the load-pull system and reduce its spurious contribution to signal quality impairment.

Maximizing modulation bandwidth can be addressed by two approaches. The first approach considers various modulation bandwidth definitions and their measurement, treated presently. The second approach involves the optimum partitioning of impedance transformation between the tuner and test-fixture, or identification of a maximum tuner $VSWR$, to ensure the overall load-pull system operates within a constant group delay regime over the instantaneous modulation deviation; this approach can be implemented with the methods of Chapter 6, on matching network design.

Definition of modulation bandwidth is arbitrary, but ultimately derives from quantitatively assessing the impairment of an arbitrary signal whose maximum instantaneous frequency or phase deviation is exposed to a substantial group delay deviation from nominal. In this context, a definition of modulation bandwidth should prescribe both the modulation and an explicit impairment metric that is repeatable, consistent, and easy to capture. The most common manifestation of impairment due to limited modulation bandwidth is the well-known phenomena of spectral asymmetry due to the odd-symmetry property of a linear time-invariant network, expressed as $H(-j\omega) = -H(j\omega)$ [3]. This phenomena is easily captured by applying a discrete two-tone signal of tone-spacing Δf and expanding the tone-spacing until the target asymmetry is reached, this tone-spacing then being defined as the modulation bandwidth.[3]

In assessing modulation bandwidth of the electromechanical tuner, consider its electrical equivalence to a single-section impedance transformer with a dynamically adjustable characteristic impedance. Since modulation bandwidth corresponds to the constant group-delay region of the input impedance of this two-port, the modulation bandwidth is equal to the region over which group delay is approximately constant, expressed as

$$\tau_g = -\frac{1}{2\pi} \frac{\partial \, arg \, \{Z_{in}\}}{\partial \omega} \approx constant \qquad (2.6)$$

and illustrated in Figure 2.14, where the modulation bandwidth spans the approximately linear phase range of the input impedance.

Modulation bandwidth of the electromechanical tuner can be measured by applying a two-tone signal and varying the tone-spacing, Δf, to a point outside the linear phase range of Figure 2.14 where the IM asymmetry exceeds an arbitrary value, commonly 1 dB. Following a reference measurement performed at 50 Ω, which yields zero asymmetry, increasing tuner *VSWR* settings enable a plot of asymmetry versus Δf. The data yields the tuner modulation bandwidth for an associated *VSWR* and can be used to for impedance transformation partition

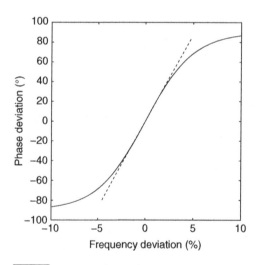

Figure 2.14 Phase response of tuner input impedance around its nominal center frequency illustrating phase nonlinearity representative of nonconstant group delay that induces IM and ACPR asymmetry. The dashed line is tangent to the approximately constant group delay region, which is a function of tuner *VSWR*.

---

3 A more rigorous method of assessing modulation bandwidth is suggested, e.g. the Noori method, when the DUT itself is suspected of contributing to modulation bandwidth, as often occurs with high-power laterally diffused metal oxide semiconductor (LDMOS) transistors with output pre-matching [4]. The DC blocking capacitor, used in the shunt inductor matching element, can be large enough to impair modulation bandwidth, or VBW, when approaching 150–200 MHz instantaneous modulation deviation commonly required for third-order DPD.

optimization between the tuner and pre-matching test-fixture to optimize overall transparency of the load-pull system to signal impairment [5]. Modern electromechanical tuners can exhibit a modulation bandwidth in excess of well over 100 MHz at maximum *VSWR* and can approach 300 MHz for moderate *VSWR*.

2.2.4 Tuner Insertion Loss

Next to *VSWR* range and operating bandwidth, insertion loss is the most important parameter in assessing electromechanical tuner performance, as it places an upper limit on *VSWR* and power capability, as well as establishing necessary driver PA power. Tuner loss also influences the optimal allocation of impedance transformation between the tuner and pre-matching network.

Intrinsic tuner loss, distinct from transition loss, is primarily due to conductor losses in the form of circulating currents coupled from the center conductor surface to ground via the probe. To reduce intrinsic tuner insertion loss, surface treatment is applied to those area where current density is highest, usually by anodizing and silver-plating. Use of low-loss transitions plays a central role in reducing overall insertion loss of the tuner; in extreme cases the transition bead is susceptible to deformation or permanent damage due to heat.

Insertion loss as defined for quantifying tuner loss is a measure of power lost within the tuner due to exclusively to heat, as distinct from mismatch loss or reflection loss contributions to total insertion loss. Heating loss is an intrinsic property of a two-port and is uniquely represented by its *s*-parameters as

$$IL = \frac{|S_{21}|^2}{1 - |S_{11}|^2} \tag{2.7}$$

where the term in the denominator compensates for mismatch loss at port 1. Equation (2.7) is commonly referred to as available loss.

Because circulating current to ground increases linearly as tuner *VSWR* increases, tuner (heating) loss increases quadratically. Tuner loss at maximum *VSWR* will determine what, if any, pre-matching is necessary to optimally allocate not only matching but to ensure the tuner operates within its specified root mean square (RMS) and peak power limits. Figure 2.15 illustrates a typical wireless band tuner loss trajectory versus the magnitude of the reflection coefficient, illustrating its rapid increase near maximum mismatch. From the trajectory four important conclusions are inferred relevant to load-pull system configuration and impedance transformation allocation

- The insertion loss of the tuner at 50 Ω is approximately 0.1 dB.
- As a load tuner, the maximum tuner *VSWR* of 100 : 1 (gamma magnitude of 0.98) yields an available loss 8.3 dB, implying that without pre-matching, nearly

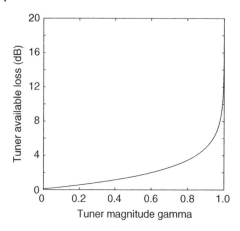

Figure 2.15 Tuner available (heating) loss in dB versus the magnitude of the reflection coefficient for the side facing the DUT.

90% of the incident power from the DUT is absorbed by the tuner versus any padding on the load side.

- As a source tuner, the maximum tuner *VSWR* of 100 : 1 (gamma magnitude of 0.98) yields an available loss 8.3 dB, implying that without pre-matching, the load-pull system driver PA will require nearly 10 dB of additional power capability.
- For a tuner *VSWR* of 10 : 1 and pre-matching network *VSWR* of 10 : 1, geometrically allocated, the available loss drops to less than 3 dB, implying the tuner will likely be within its rated peak and RMS power capability as well substantially reducing driver PA requirements.

2.2.5 Power Capability

Electric field breakdown and localized Joule heating are the two physical mechanisms establishing an upper bound on peak power and average power capability of an electromechanical tuner, respectively. Both breakdown mechanisms are typically avoided by adoption of pre-matching networks to reduce the peak electric field and circulating currents within the tuner. Low quality adapters present at the tuner and fixture interface are often the source of both electric field breakdown and Joule heating, which is easily addressed by adopting high-quality adapters.

Electric field breakdown commonly occurs in load-pull with simultaneous high peak–average signals and high tuner *VSWR*, since these create conditions conducive to electrical breakdown of air, which occurs at approximately 31 kV/cm. Corona discharge is the physical process responsible for electric field breakdown, and can be envisioned with Figure 2.1 with the probe near the center conductor surface and the concomitant increase in electric field intensity. While not necessarily catastrophic to the tuner, unless the same probe and carriage locations are used repeatedly, the intense heat of the arc may cause oxidation, leaving permanent pock marks on the center conductor surface and probe, changing their physical constitution. Since corona discharge produces ozone, it can be easily detected from its distinguishing odor. Modern load-pull tuners are capable of multi-hundred watt load-pull by judicious choice of pre-matching fixture design coupled with dynamic pre-matching tuner technology that reduces peak electric field and heating by allocating matching over several sections.

Under high *VSWR* conditions, the probe is in proximity to the center conductor, increasing circulating current to ground. The increase in circulating current, for constant voltage, is the physical basis of low-pass impedance transformation, creating substantial heat due to metal loss. Heat removal is particularly acute for the probe due to its low thermal mass and small area, reducing both conductive and convective heat transfer. Tuner damage due to heat is seldom an issue, unless it is operating near its maximum *VSWR*. A far more significant and nefarious problem with self-heating is temporary expansion of internal tuner components, particularly longitudinal compression of the center conductor, causing it to bow or sag. This compression induces a transverse displacement of the center conductor with respect to the probe, manifested as a systematic, but uncalibrated, error in its *s*-parameters.

A rigorous method to assess an acceptable upper limit on tuner RMS power due to self-heating is illustrated by Figure 2.16. The first tuner acts as a pre-match for the high-power driver PA, with the second tuner acting as a virtual load generating an impedance in the neighborhood of R_{min} of Figure 2.3, matched to the driver tuner. Two directional couplers, separated by an isolator to enhance directivity due to the large mismatch present at the input, sample the reflected power to measure the input impedance of the tuner cascade versus available source power. After allowing thermal steady-state to be achieved at each load power, impedance deviation from room-temperature can be assessed to determine the maximum allowable change at a target RMS power. For a typical 8 GHz tuner with substantial thermal mass, being driven with 200 W RMS, the magnitude

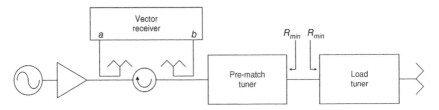

Figure 2.16 Configuration to evaluate maximum tuner RMS power capability for a specified change in tuner impedance due to self-heating. Tuner impedance R_{min} refers to lowest impedance state of each tuner, as defined in Figure 2.3.

Table 2.1 Maximum PEP and RMS power ratings by tuner frequency and connector style, subject to the constraints listed in the text.

Tuner frequency	Connector style	Maximum PEP (W)	Maximum RMS power (W)
200 MHz to 7.0 GHz	7–16 DIN coaxial	1 kW	500
400 MHz to 12.0 GHz	Type N coaxial	500 W	150
800 MHz to 18.0 GHz	7 mm APC coaxial	500 W	150
800 MHz to 26.5 GHz	3.5 mm coaxial	200 W	75
800 MHz to 50.0 GHz	2.4 mm coaxial	75 W	25
75–110 GHz	WR-10 Ag-plated waveguide	1 kW	100

a) PEP and RMS power are specified under mismatch condition versus 50 Ω and assume high-quality tuner transitions free from metallic spurs and other defects.

of the reflection coefficient change looking into the load tuner can exceed 1%, equivalent to a repeatability of −20 dB.

Table 2.1 summarizes peak and RMS power capability of tuners by frequency and connector style. Throughout the table, it is assumed that high quality connectors used. It is also assumed pre-matching is deployed, so that the tuner available loss is in the neighborhood of 1 dB or so. Tuner operation with the probe on the outer-ring represents a maximum available loss condition in which the tuner is exposed to full dissipation of the RF signal, as well impaired repeatability, and is not advised.

2.2.6 Vector Repeatability

Vector repeatability of the passive electromechanical is a scalar measure, usually given in decibels, of the vector displacement between two nominally identically impedance states. Because repeatability tends to degrade at maximum *VSWR*, it is often specified at both 50 Ω and at a specified $VSWR_{reference}$ that is adopted consistently for each repeatability measurement. A well-designed electromechanical tuner will exhibit a vector repeatability better than of −50 dB at 50 : 1 *VSWR* from HF to mm-wave, for modest *VSWR*. At maximum *VSWR*, usually defined as the associated *VSWR* at the outer-ring, a vector repeatability better than −40 dB is typical.

Figure 2.17 provides a graphical definition of vector repeatability, illustrating it is the vector difference between the reference impedance state and the test impedance state, though the result is often simplified to report the magnitude of the difference vector. Nonideal mechanical phenomena, such as backlash, introduce a hysteresis and path dependence to probe and carriage displacement,

Figure 2.17 Graphical illustration of tuner vector repeatability.

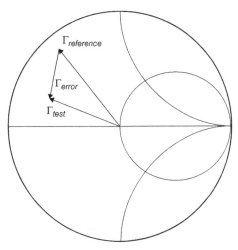

introducing vector memory component to the repeatability metric. For this reason, vector repeatability is often characterized by an established reference impedance state and test state, and a specified intermediate probe and carriage trajectory, for example an initial impedance state of 50 Ω, movement to $\Gamma_{reference}$, movement back to 50 Ω, and final movement to Γ_{test}. An alternative is to perform the characterization as an average of several different intermediate probe and carriage trajectories to the final test impedance state, Γ_{test}, from $\Gamma_{reference}$, without regards to each intermediate trajectory.

Measurement of vector repeatability requires a well-calibrated VNA with clean and gauged connectors. It is recommended that the tuner port facing the DUT be connected directly to the VNA, eliminating the test-port cable, and that the VNA source-match be better than the repeatability being measured. A source match less than the tuner repeatability attempting to be measured substantially impairs the accuracy of the repeatability measurement. Vector repeatability in scalar form is defined as

$$|\Gamma_{error}| = |\Gamma_{reference} - \Gamma_{test}| \tag{2.8}$$

where each variable is defined in Figure 2.17.

2.2.7 Impedance State Resolution and Uniformity

Impedance state resolution establishes the ability to capture optimum, or target, DUT performance by providing the required reflection coefficient foretasted from contour data. Insufficient resolution means that the tuner cannot physically synthesize the required reflection coefficient to deliver the performance predicted from contour data. Similarly, tuner uniformity establishes the relative concentration of reflection coefficient states on the interior of the Smith chart, with both tuner limitations and impedance transforming test-fixtures contributing to concentration of reflection coefficient states to preferred regions on the interior of the Smith chart.

The electromechanical tuner employs stepper motors with sufficient physical resolution to provide reflection coefficient magnitude displacement better than 0.005 and phase displacement better than 0.1° at its highest operating frequency. This level of resolution is achieved using two-dimensional interpolation and does not generally require calibration at all reflection coefficient states. Therefore, tuner calibration using several hundred reflection coefficient states is sufficient to generate literally millions of interpolated reflection coefficient states. For applications that require the highest fidelity and repeatability, most load-pull software packages provide the option to snap to measured reflection coefficient states, thereby eliminating interpolation error.

Impedance transformation, such as that generated by quarter-wave prematching, concentrates reflection coefficient states near the characteristic impedance of the transformation.[4] This concentration, with respect to 50 Ω, is generally not a serious limitation since most data of interest are concentrated in the region of the transformation.

2.2.8 Factors Influencing Tuner Speed

Three primary factors establish passive electromechanical impedance synthesis speed. The physical length of the tuner, which sets the lower frequency boundary, affects speed simply by the additional time required for the carriage to create a given angular displacement around the center of the Smith chart. Stepper motor speed affects speed in a similar fashion insofar as a faster motor will traverse a specified distance faster than a slower stepper motor.

Each probe and carriage displacement pair represent a unique impedance for constant frequency. How the probe and carriage are instructed to move to the next requested impedance state plays a central role in establishing the overall speed of the electromechanical tuner, being essentially a form of the well-known traveling salesman problem [6]. The translation of requested tuner impedance states into a time-optimal transition problem will substantially reduce load-pull time since it avoids multiple repetitive probe and carriage displacements, especially the former, which is generally slower due to its larger inertia. Some commercial load-pull software packages offer a software option to compute a minimum-time solution versus the standard approach, which simply moves the probe and carriage the requested displacement based on the ordering from the user.

---

4 This concentration is actually illusory, however, being an artifact of the bilinear transformation used to generate the Smith chart. Simply resetting the reference impedance from 50 Ω to the characteristic impedance of the transformation will restore apparent uniformity.

2.2.9 The Slab-Line to Coaxial Transition

To synthesize an arbitrary impedance, the electromechanical tuner exploits constructive and destructive interference present on a mismatched transmission line, resulting in periodic voltage and current maxima. The interface between the transmission line internal to the tuner and its external environment, typically a slab line and coaxial line, is realized by a precision transition that transports energy between media with minimum return loss and insertion loss, and may be exposed to voltage and current maxima. This results in electrical and thermal stress, similar to that of probe proximity to the surface of the center conductor. The capacity of the transition to rapidly and efficiently dissipate heat is a limiting factor in establishing the maximum *VSWR* and maximum power capability of the tuner.

To minimize the influence of standing waves on loss, modern electromechanical tuners therefore employ high-performance slab-line to coaxial transitions, free of mechanical defects and metal spurs, with highly polished gold or silver surface finish. Axially symmetric center conductor support is maintained by precision low-loss precision dielectric beads. A low quality coaxial adapter, or often an inconspicuously damaged adapter, will contribute substantial insertion loss. In addition to reducing peak tuner *VSWR*, the presence of a voltage or current standing wave maxima within such an adapter leads to substantial, and possibly harmful, surface temperatures under high-power load-pull. For this reason, each adapter used in a load-pull configuration should be visually inspected, gauged, and verified with a VNA prior to use to ensure its return loss and insertion loss support the anticipated operating conditions.

2.3 Advanced Considerations in Impedance Synthesis

2.3.1 Independent Harmonic Impedance Synthesis

Several methods are available for independent harmonic impedance synthesis. First-generation methods were largely *ad hoc* attempts that resulted in substantial ancillary impairment of primary performance factors, particularly maximum *VSWR*, operating bandwidth, and repeatability. One of the first commercial harmonic load-pull methods employed frequency multiplexers to provide isolation between multiple tuners in parallel, with each tuner assigned to a harmonic, and one to the fundamental. Though this method was intuitive and provided excellent repeatability, it resulted in a substantial *VSWR* contraction due to multiplexer insertion loss. An alternative method relied on insertion of multiple secondary probes resonant at a particular harmonic, usually the second and third. While this method offered little reduction in *VSWR*, it required adaptive tuning at the fundamental to compensate for the poor isolation between the fundamental and

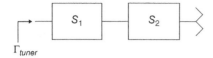

Γ_{tuner}

Figure 2.18 Two-port representation of two carriage–probe pairs cascaded on a shared transmission-line, similar to the configuration illustrated by Figure 1.1.

harmonic probes. Both of these methods suffered from relatively narrow operating bandwidth, typically on the order of 10% of the operating frequency, and therefore required a collection of hardware for a typical load-pull system, as well as frequent calibration as the operating frequency was changed.

A clever method of harmonic impedance synthesis exploits frequency dispersion to *a priori* established distinct probe and carriage positions, of a plurality of probe–carriage pairs, to unique impedances defined by arbitrary frequencies, not necessarily harmonically related [7]. To understand the underlying theory of the this method, consider adjacent probe–carriage pairs in proximity, situated on a shared transmission line enclosed in a common housing, similar to that shown in Figure 2.4. This configuration can be represented electrically as a cascade of two distinct two-port networks, as shown in Figure 2.18, where S_1 represents the s-parameters of carriage–probe pair one from where the probe is aligned longitudinally on the slab-line to the reference-plane, defined at the end of the tuner closest to carriage–probe pair one.

Similarly, S_2 represents the s-parameters of carriage–probe pair two from where the probe is aligned longitudinally on the slab-line to the reference-plane, defined at the end of the tuner closest to carriage–probe pair one. It is important to note that the s-parameter definitions attached to each carriage–probe pair assume the adjacent probe is fully retracted so that the slab-line appears as a homogeneous 50 Ω transmission line in either direction.

Now the slab-line is represented as a lossless transmission-line with frequency-dependent phase velocity, $v_p(j\omega)$. Assuming d_1 to be the displacement of carriage–probe pair one with respect to the tuner reference-plane and d_2 to be the displacement of carriage–probe pair two with respect to the tuner reference-plane, the phase retardation from carriage–probe one to the tuner reference-plane is

$$\theta_1 = \frac{\omega_1 d_1}{v_p(j\omega)} \tag{2.9}$$

and the phase retardation from carriage–probe pair two to the tuner reference-plane is

$$\theta_2 = \frac{\omega_1 d_2}{v_p(j\omega)} \tag{2.10}$$

The phase retardation terms given by Eqs. (2.9) and (2.10) reset the reference-plane of each carriage–probe pair to the tuner reference-plane. The reflection coefficient of the cascade of carriage–probe pairs becomes

$$\Gamma_{tuner}(j\omega) = [S_{11}]_1 e^{-j2\theta_1} + \frac{[S_{21}]_1 [S_{12}]_1 [S_{11}]_2 e^{-j2\theta_2}}{1 - [S_{22}]_1 [S_{11}]_2 e^{-j2\theta_2}} \tag{2.11}$$

where the first term represents the contribution due to the first carriage–probe pair, the second term represents the second carriage–probe pair, and the brackets denote the frequency-dependent s-parameter matrices of the first and second two-ports of Figure 2.18.

To illustrate the harmonic impedance synthesis algorithm, assume each carriage–probe pair is calibrated at frequencies ω_1 and $\omega_2 = 2\omega_1$. Linear interpolation on Eq. (2.11) attaches to each element of the set of physical carriage–probe pair combinations an s-parameter matrix at ω_1 and ω_2, yielding a combinatorial array for subsequent application of constrained optimization, with the objective function composed of the target reflection coefficient states. Define now Γ_1 and Γ_2 as the target reflection coefficients at the tuner reference-plane at fundamental ω_1 and second-harmonic ω_2, respectively. Consider next a circular region centered at Γ_1, of radius $|\delta|$, at radian frequency ω_1. Following these definitions, Eq. (2.11) as an explicit function of ω_1 and ω_2 is

$$\Gamma_{tuner}(j\omega) = \begin{cases} \Gamma_1 + \delta e^{j\psi}, & \omega = \omega_1 \\ \Gamma_2, & \omega = \omega_2 \end{cases} \tag{2.12}$$

where $|\delta|$ is usually chosen to be less than the equivalent tuner repeatability at radian frequency ω_1 and ψ is composed of a sufficient number of phase steps to span 0–2π radians at the tuner reference-plane. Equation (2.12) is represented graphically on the Smith chart of Figure 2.19, showing the fundamental and second harmonic reflection coefficient vectors and the circular regions swept by $\delta e^{j\psi}$, shown enlarged for clarity.

To properly interpret the underling physics embodied by Eq. (2.12), it is crucial to recognize that each term of Eq. (2.12) does not represent the contribution of an individual carriage–probe pair, but rather the composite tuner reflection coefficient produced by an *a priori* established combination of physical displacements of the two carriage–probe pairs, subject to the constraint $\delta e^{j\psi}$ at ω_1. Each of these reflection coefficient states produced by this composite subsequently map to an identical number of reflection coefficient states at ω_2, distributed over the entire Smith chart. It is this expansionary mapping process, due to the frequency dispersion specified by Eqs. (2.9) and (2.10), that produces a set of second-harmonic reflection coefficients over the entire Smith chart due to small change in the reflection coefficient constrained by $\delta e^{j\psi}$ at ω_1.

It is also important to recognize the circular region of reflection coefficients, at radian frequency ω_1, is not due to a prescribed physical carriage–probe displacement. Rather, it represents the set of all carriage–probe displacement pairs

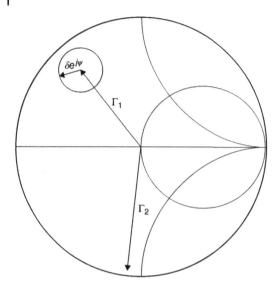

Figure 2.19 Fundamental and harmonic impedance vectors, Γ_1 and Γ_2, respectively, and the $\delta e^{j\psi}$ vector sweeping out the constraint circle at ω_1. The circle is shown several orders of magnitude larger than its typical radius to illustrate its location on Γ_1.

that produce $\Gamma_{tuner}(j\omega_1)$ subject to the constraint $\delta e^{j\psi}$. There are in theory an infinite number of combinations of the carriage–probe pairs that produce distinct fundamental and second-harmonic reflection coefficients, even within the circle defined by $\delta e^{j\psi}$, though in practice the combinations are limited to millions due to finite resolution of the carriage–probe stepper motors.

To illustrate the mechanics of the present harmonic impedance synthesis method, consider Figure 2.20. Shown on the interior of the circle defined by $\delta e^{j\psi}$ are several fundamental impedances states, whose maximum displacement from Γ_1, the desired fundamental impedance, establish an isolation proxy with the desired second-harmonic impedance. This fundamental impedance perturbation is balanced against sufficient flexibility to produce the required uniformity, distribution, and maximum value of the second-harmonic impedance states.

A reasonable initial constraint is $|\delta| \leq 0.0001|\Gamma_1|$, which is equivalent approximately to -40 dB tuner repeatability. Each objective function term can be optimized so that, for example, specific weighting could be placed on high-$VSWR$ fundamental load-pull while simultaneously holding the second-harmonic constant. The weighting terms and $|\delta|$ collectively enable sufficient flexibility to adapt multi-probe impedance generation to various classes of impedance synthesis problems, including harmonic impedance synthesis, minimum group-delay synthesis, and the important class of frequency-agile dynamic pre-matching methods, next treated.

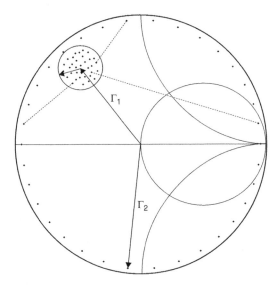

Figure 2.20 Fundamental and harmonic impedance vectors, Γ_1 and Γ_2, respectively, and the $\delta e^{j\psi}$ vector sweeping out the constraint circle at ω_1. The circle is shown several orders of magnitude larger than its typical radius to illustrate its location on Γ_1. The dashed lines show the correspondence between the fundamental states on the interior of $\delta e^{j\psi}$ and the second-harmonic states on the boundary of the Smith chart. This expansionary mapping process, due to frequency dispersion, is the basis of the multi-probe method of harmonic load-pull with electromechanical tuners.

2.3.2 Sub-1 Ω Impedance Synthesis

Since its rigorous application to load-pull, distributed pre-matching has been widely adopted for sub-1 Ω impedance synthesis [8]. Distributed matching can be applied to narrow-band applications with a quarter-wave line, octave bandwidth applications with a multi-section Chebyshev transformer, and decade wide-band applications with the Hecken taper [2, 9]. Nevertheless, there are applications where multi-decade bandwidth pre-matching is necessary or pre-matching may be physically impractical, in which case frequency agile pre-matching is required. Typical examples include wideband unit-cell model verification or on-wafer mm-wave applications with substantial probe loss.

Frequency agile dynamic pre-matching with the electromechanical tuner can be implemented by one of four methods. The first method relies on corrugated-probe technology, in which the probe is designed as a multi-element periodic structure to synthesize a *VSWR* larger than what is possible with one probe alone [10]. The corrugated probe acts as a multi-section impedance transformer that is capable of 200 : 1 *VSWR* at common the wireless bands spanning 800 MHz to 6 GHz. While this method is relatively simple, and offers low cost, it suffers from reduced repeatability and characterization difficulty, due to the extremely high mismatch presented to the VNA during tuner calibration.

The second method of pre-matching is done by using a first probe as a pre-matching element and a second probe as an impedance tuning element, each colocated on a common slab-line as shown in Figure 2.4. This approach is similar to quarter-wave pre-matching, with the pre-matching element equivalent a transmission-line whose quarter-wave length and characteristic impedance are continuously variable.

The multi-probe method effects frequency-agile dynamic pre-matching by identifying those carriage–probe pairs producing collinear Γ_1 and $\delta e^{j\psi}$ aligned radially outward from the center of the Smith chart. Under this optimization criteria, two carriage–probe pairs are calibrated at ω_1 and those collinear carriage–probe pairs in phase with one another substantially extend the maximum tuner *VSWR*. A substantial benefit of the multi-probe method is its ability to effect various classes of impedance synthesis in one compact enclosure, replacing what was several individual components, including frequency multiplexers for harmonic load-pull and distributed pre-matching elements for sub-1 Ω load-pull.

Active and hybrid impedance synthesis is ideally suited for mm-wave on-wafer load-pull where probe losses render it physically impossible to achieve a practical *VSWR* using an electromechanical tuner. Figure 2.21 illustrates *VSWR* at the probe reference-plane versus tuner *VSWR* for various probe insertion losses. Even for a state-of-the-art probe insertion loss of 0.8 dB at 77 GHz, a tuner with infinite *VSWR* yields a gamma magnitude of 0.68 at the probe-tip reference-plane. Because active and hybrid synthesis can synthesize a reflection coefficient magnitude greater than unity, probe losses are entirely neutralized, yielding an impedance at the probe reference-plane anywhere on the interior of the Smith chart, as well as on its boundary, or even its exterior, to compensate for probe and cable loss.

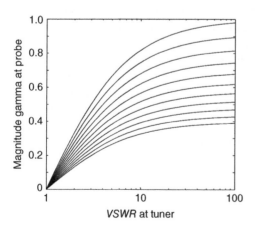

Figure 2.21 Gamma magnitude at the probe-tip reference versus tuner *VSWR* for probe insertion loss from 0 to 2.0 dB in 0.2 dB increments.

2.4 Closing Remarks

The electromechanical tuner is the foundation of modern load-pull. Its flexibility and evolutionary maturity serve a broad class of applications, spanning impedance range, power capability, operating bandwidth, and independent harmonic control, while providing superior repeatability and accuracy with reasonable measurement speed. Supplemental factors, such as tuner insertion loss, vector repeatability, and impedance state uniformity, are suitable for the vast majority of applications.

In contrast to first-generation electromechanical tuners of the early 1990s, second- and third-generation impedance synthesis architectures are capable of independent harmonic control and frequency-agile sub-1 Ω impedance synthesis. The third-generation of impedance synthesis includes several commercially practical active architectures, including active-loop and active-injection. Though active-loop is, and will, remain expensive for high-power applications, active-injection provides substantial improvement in speed and resolution, and the unique capability for fixture- and probe-loss neutralization.

References

1 Takayama, I. (1976). A new method of active load-pull. IEEE International Microwave Symposium.

2 Tsironis, C. (2005). *Precision Measurements for the 21st Century*. Montreal, QC Canada: Focus Microwaves, Inc.

3 Sevic, J.F. and Steer, M.B. (1998). A novel envelope termination method for ACPR optimization of RF and microwave power amplifiers. IEEE Transactions on Microwave Theory and Techniques Symposium Digest.

4 Noori, B. (2004). A simple method of calculating VBW in-fixture. IEEE Radio and Wireless Symposium.

5 Sevic, J.F. (2005). Tuner Instantaneous Bandwidth, Linear Network Distortion, and IM and ACPR Load-Pull. *Application Note 5C-063*. Maury Microwave Company.

6 Applegate, D.L. and Bixby, R.E. (2006). *The Traveling Salesman Problem: A Computational Study*. Princeton, NJ: Princeton University Press.

7 Tsironis, C. (2004). Triple-probe automatic slide-screw load-pull tuner and method. US Patent 7, 135, 941, Tech. Rep.

8 Sevic, J.F. (1996). A rigorous two-tier fixture calibration method for low impedance load-pull. Automatic RF Techniques Group Conference.

9 Noori, B. and Sevic, J.F. (2006). Load-pull: accuracy versus reality. IEEE Radio and Wireless Symposium.

10 Sevic, J.F. and Simpson, G.M. (2005). *High-Gamma Tuner Product and Data Note*. Maury Microwave Company.

3

Load-Pull System Architecture and Verification

RF and microwave impedance synthesis, both passive and active, is the systematic design of variation and control of the relative magnitude and phase of traveling waves. Because loss and distributed network effects are considerable at RF and microwave frequencies, a substantial reduction in matching range and introduction of other impairments can result between the intrinsic load-pull system and device under test (DUT) reference-plane. This requires that particular attention be given to the system architecture embedding the DUT within the load-pull system to ensure that minimal impairment is introduced to impedance range, operating and modulation bandwidth, measurement speed, and power capability.

System architecture also establishes repeatability, stability, and accuracy of the measured data, with overall system performance being assessed by rigorous verification, similar to the method employed for the VNA. By relying on the traceability of the VNA calibration standards to an arbitrary reference impedance, the repeatability performance of the load-pull system can be precisely established and related to the effect on measured parameters. This is useful in its own right to ensuring precise results, but equally important for quantifying and assessing long-term error performance and resolving slow changes in performance due to aging, drift, and other physical factors.

The present chapter is an introduction to the architecture of load-pull, with emphasis placed on passive impedance synthesis. The general architecture is analyzed, supported by a thorough description of the performance requirements of each section of the system. Measurement of large-signal input impedance is presented, as are load-pull of amplitude to amplitude conversion distortion (AM–AM) and amplitude to phase conversion (AM–PM) and dynamic range optimization. System accuracy verification is given rigorous treatment by introducing the ΔG_T method, a widely adopted approach to verifying load-pull system repeatability performance and its effect on other parameters, such as power-added efficiency (PAE) [1–3].

The Load-Pull Method of RF and Microwave Power Amplifier Design, First Edition. John F. Sevic.
© 2020 John Wiley & Sons, Inc. Published 2020 by John Wiley & Sons, Inc.

3.1 Load-Pull System Architecture

Architecture establishes matching range, frequency range, power capability, and ease-of-use, in addition to directly influencing repeatability and accuracy. A well-designed architecture greatly contributes to enhanced performance, repeatability, and accuracy. More importantly, a well-designed architecture provides confidence in data and extends system calibration and verification intervals.

The fundamental reliance of load-pull on precise de-embedding and DUT reference-plane shifting, and the subsequent assumption of stability and repeatability, stipulates use of high-performance hardware with minimum sensitivity to external perturbation or influence. This introduces potential tension in achieving maximum performance over several parameters simultaneously, such as matching range and frequency response. As such, a specific architecture may offer maximum matching range at the expense of frequency response or, similarly, offer broadband operation at the expense of matching range. Figure 3.1 shows a contemporary high-performance on-wafer load-pull system, illustrating the various types of measurement hardware necessary for rigorous characterization of transistor performance.

3.1.1 Load-Pull System Block Diagram

Figure 3.2 illustrates a generic load-pull system architecture based on passive impedance synthesis, though the principles of the present chapter apply

Figure 3.1 Contemporary high-performance on-wafer load-pull system.
Source: Reproduced with permission of Maury Microwave, Inc.

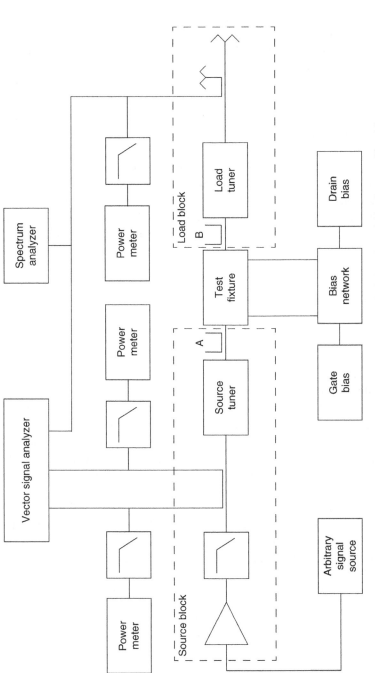

Figure 3.2 Generic load-pull system illustrating various functional blocks and their location within the overall architecture.

equally well to active impedance synthesis. Fundamental and harmonic passive impedance synthesis is assumed to obtain from the multiple-probe method introduced in Chapter 2, with, subharmonic load-pull, for memory optimization, implemented via the DC bias ports [4]. Hence, from an architecture perspective, the impedance synthesis method is embodied by the source and load tuners shown.

The choice of a specific architecture is largely driven by the DUT parameters under consideration. These parameters are broadly classified as independent, dependent, and derived. The source and load blocks are associated with conditioning of independent parameters, from which dependent and derived parameters are sampled and extracted, respectively. Independent parameters are the conditions exposed to the DUT that uniquely establish its operating state, from which all other operating parameters obtain. Independent parameters are most often, but not always, the impedance presented to the DUT and associated bias conditions.

3.1.2 Source and Load Blocks

The source and load blocks of Figure 3.2 must immunize DUT performance from impairments due to the presence of the load-pull system itself, and simultaneously sample independent and dependent signals with minimum impairment. The signal source is often followed by modest padding and an amplification stage capable of providing sufficient available source power at the DUT reference plane. The source directional coupler samples incident and reflected source power, which in general will include phase measurement, particularly for DUT large-signal input impedance measurement, required for measurement of true PAE. Similarly, the load directional coupler samples incident load power, with coupling chosen to match expected load power with the power capability of the power sensor or signal analyzer.

Figure 3.3 illustrates in expanded scale the impedance seen looking back from the DUT, toward the source tuner and source block. The DUT source impedance is thus seen to be function of not only the source tuner, but the impedance terminating the source tuner, which itself is a function concatenated *s*-parameters of the coupler, padding, and reference power amplifier (PA).[1]

The impedance seen looking into the source tuner, from the DUT side, is

$$\Gamma_{Source\text{-}Tuner} = S_{11} + \frac{S_{21}S_{12}\Gamma_{Source\text{-}Block}}{1 - S_{22}\Gamma_{Source\text{-}Block}} \tag{3.1}$$

---

1 The source bias tee is assumed to be on the DUT test-fixture. If the bias tee is placed within the source block, then its *s*-parameters too must be accounted for.

Figure 3.3 Smith chart illustrating vector dependence of DUT source impedance on source tuner, the source block, and the source terminating impedance, composed of the signal source and possibly a reference PA.

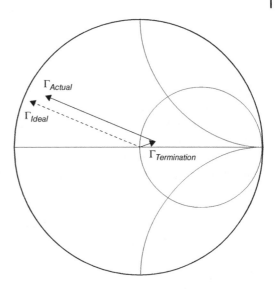

where the S_{ij} are the s-parameters of the source tuner. The impedance terminating the source tuner is best represented by the concatenation of the t-parameters of the elements composing the source block

$$T_{Source\text{-}Block} = T_{S1} T_{S2} T_{S3} T_{S4} \tag{3.2}$$

where the T_{Si} include as a minimum the source coupler and the terminating impedance presented by the signal source or reference PA. As can be seen from the present analysis, substantial deviation from the apparent tuner impedance, presented to the DUT, will occur with nonideal source block elements. Similarly, impairment introduced by the source block that is out-of-band, particularly at the envelope frequency and harmonics must be accounted for, and often deliberately controlled, to reduce potential for oscillation and introduction of memory.

Immunization of DUT performance from nonideal effects of the source block begins with estimating the required available source power at the DUT reference-plane. Because a passive tuner exhibits substantial loss at extreme mismatch, the expected mismatch must be identified, with the associated insertion loss presented by source tuner identified. Given the required available source power and source tuner loss, and the available source power of the signal source, a decision is made on the necessity of a reference PA, whose own performance must provide adequate drive without introduction of its own impairments. Reference PA signal impairment is particularly a concern when assessing DUT linearity performance, such as adjacent channel power ratio (ACPR), and harmonic generation, whose effects are removed by the addition of a low-pass filter.

Following reference PA selection, couplers and pads are chosen. Couplers are generally of the dual directional coupler variety, with 20 dB coupling being a reasonable compromise in dynamic range. The role of the pad, though seemingly innocuous, is vital in immunization by reducing the effect of the reference PA output mismatch, seldom better than 2 : 1. The pad also plays an important role in setting out-of-band impedance, and it is for this reason that use of a circulator is not recommended in load-pull, due to their extreme out-of-band mismatch. Obviously a trade-off exists in padding and reference PA power, with 3 dB a reasonable choice for padding the source block and source tuner and 6 dB as reasonable choice for padding the signal source and reference PA.

Optimization of the source block architecture component properties, particularly reference PA power, coupling ratios, and pad values, is followed by consideration of signal sampling for measurement. The two primary independent signals sampled at the source block are incident power and reflected power, each of which must be reset from the physical reference-plane to the measurement reference-plane. Further, in some applications, the phase of the incident and reflected power waves must be measured, as shown by the presence of the splitter on incident port of the source coupler, followed by a complex signal analyzer.

Finally, low-pass filters are used to reduce the effect of harmonics on apparent measured power introduced at the power sensor, whose effect is quantified in Figure 3.4. From this plot, it is seen that better than 20 dB harmonic rejection should be considered the minimum required to obtain an accurate PAE measurement.

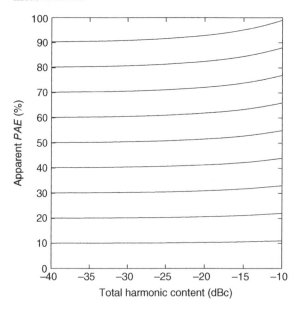

Figure 3.4 Apparent PAE versus total harmonic content with respect to fundamental power parametrized to actual PAE to assess optimum low-pass filtering requirements.

Once the source block architecture is settled, it remains to properly characterize it for de-embedding. It is generally good practice to measure the s-parameters of each element comprising the source-block to identify potential problems, for example due to mechanical damage of the connector or internal damage due to exceeding maximum power limits. In principle, the s-parameter matrix of each element can be used from this step, and cascaded to form a composite s-parameter matrix, though treating the assembled source-block as one element reduces the influence of connector repeatability between elements.

The source-block is composed of multiple ports, specifically the incident signal at the reference PA side, the reflected signal at the source tuner side, and the coupled ports. These s-parameters describing the behavior of these ports are measured with a VNA, calibrated by TRL, over a frequency range spanning the modulation frequency to at least the third harmonic. Though load-pull tools often include a facility for interpolation of s-parameters between measured frequencies, it is good engineering practice to include the specific load-pull measurement frequencies in the frequency list, followed by a frequency list that spans the desired overall frequency range.

Characterization of the source block is completed by measuring the output impedance of the signal source or reference PA, depending on architecture, since the transmission vector of the source block sits on the vector looking back toward the source using again a VNA set for low power.[2] This measurement assumes the output impedance is power invariant, which is often a good approximation.

The impedance seen looking into the load tuner, from the DUT side, is

$$\Gamma_{Load\text{-}Tuner} = S_{11} + \frac{S_{21}S_{12}\Gamma_{Load\text{-}Block}}{1 - S_{22}\Gamma_{Load\text{-}Block}} \tag{3.3}$$

where the S_{ij} are the s-parameters of the load tuner. The impedance terminating the load tuner is best represented by the concatenation of the t-parameters of the elements composing the load block

$$T_{Load\text{-}Block} = T_{L1}T_{L2}T_{L3}T_{L4} \tag{3.4}$$

where the T_{Li} include as a minimum the load coupler, a low-pass filter, and the terminating impedance presented by the load termination. In following with the source block, substantial deviation from the apparent tuner impedance, presented to the DUT, will occur with nonideal source block elements. Similarly, impairment introduced by the load block that is out-of-band, particularly at the envelope frequency and harmonics must be accounted for, and often deliberately controlled, to reduce potential for oscillation and introduction of memory.

---

2 This measurement requires caution when measuring the output impedance of a high-power PA, due to the potential for VNA damage.

Immunization of DUT performance from nonideal effects of the load block begins with estimating the maximum estimated load power delivered by the DUT to optimize allocation of impedance transformation that keeps the load tuner from an extreme mismatch condition, and its associated maximum peak and root mean square (RMS) power capability. Given the estimated DUT load power, load tuner loss, and maximum power capability of the power sensor (or signal analyzer), necessary padding and coupling can be calculated.

The load coupler can be generally the single directional coupler variety, with 20 dB coupling being a reasonable compromise in dynamic range and apparent attenuation. Because the power sensor generally has a larger dynamic range than the dynamic range of a signal analyzer, and padding contracts dynamic range due to a noise contribution approximately equal to its attenuation, it is good practice to sample power following the required padding. Signal quality sampling is thus done at the coupled port of the coupler, to provide apparent attenuation without dynamic range impairment. Finally, similar to the source block, the phase of the load power wave must be measured, as shown by the presence of the splitter on the coupler, followed by a complex signal analyzer.

Low-pass filtering is mandatory in the load block, due to substantial harmonic generation. Figure 3.4 illustrates this effect, illustrating that better than 20 dB harmonic rejection should be considered the minimum required to obtain an accurate PAE measurement. In some applications, it is good practice to insert a low-pass filter in front of the signal analyzer.

Once the load-block architecture is settled, it remains to properly characterize it for de-embedding, following the process described for the source block. In measuring the s-parameters of the source and load blocks, it is important to understand the convention port definitions adopted by the load-pull software being used, since the convention is not standardized.

3.1.3 Signal Synthesis and Analysis

A significant benefit of the load-pull method is the use of realistic stimulus, thus avoiding errors associated with, for example, inference of ACPR from intermodulation (IM). Therefore, it is required that the signal generator of Figure 3.2 be capable of not only single-tone continuous wave (CW) capability, but having arbitrary waveform capability with sufficient bandwidth and quantization to replicate signals such multi-carrier WCDMA, LTE, and 5G.

Signal quality measurements under load-pull similarly requires instrumentation capable of supporting the types of measurements sought. Measurement systems are required to measure RF power, DC bias, and, in many cases, provide frequency-domain and demodulation capability, for ACPR and error vector magnitude (EVM), respectively. Careful attention must be given to power

Figure 3.5 Voltage and current conventions that define the large-signal input impedance of a transistor, at a specific frequency, usually the fundamental.

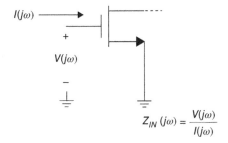

$$Z_{IN}\,(j\omega) = \frac{V(j\omega)}{I(j\omega)}$$

measurement of wide-band modulated signals by ensuring the power sensor modulation bandwidth matches the bandwidth of the stimulus and its harmonics. Thus, for example, a 20 MHz LTE stimulus requires a modulation bandwidth of at least 100 MHz to capture fifth-order spectral regrowth.

3.1.4 Large-Signal Input Impedance Measurement

It is evident that a multistage PA will have at least one interstage matching network terminated at each port by a transistor, possibly operating in its large-signal regime. Since one terminating impedance must be known to complete the interstage matching network synthesis, it is convenient to define a large-signal impedance looking into the base/gate of a transistor as the ratio of its voltage to current, at a *specific frequency*, as illustrated in Figure 3.5. While this impedance is subject to change over the large-signal operating regime of the transistor, it is nevertheless a well-posed physically meaningful quantity that can be measured easily on a load-pull system [5].

Large-signal input impedance measurement with passive load-pull requires a two-port complex receiver configured in ratio mode, as shown in the source block Figure 3.2. A calibration port is chosen, usually for measurement convenience, and a one-port SOL calibration is performed over the desired frequency range. The measurement reference plane is then reset to the DUT base/gate using the measured s-parameters of the source block. Because the large-signal input impedance is a function of frequency, and power, it is important to understand its behavior over drive, particularly for optimization under high drive, to maximize delivered power.

3.1.5 AM–AM, AM–PM, and IM Phase Measurement

Certain PA architectures, such as digital pre-distortion (DPD), require that AM–AM and AM–PM distortion be known. This is possible in load-pull by adding to the vector receiver of Figure 3.2 an additional channel that samples the DUT at the load tuner, thus yielding transducer gain magnitude and phase, which are AM–AM and AM–PM. By appropriate signal processing, memory characterization is also possible.

It has been long known that PA mixing products are in general complex, and therefore subject to constructive and destructive interference effects among cascaded stages of an RF PA [6]. In modern times this phenomena is frequently referred to as in-line pre-distortion, which can in special cases be exploited to improve the overall linearity of the PA by matching network design that deliberately controls IM phase of one stage to cancel the IM contribution of successive stages.

The phase of IM can be measured using large-signal vector network analysis capability large signal network analyzer (LSNA) [7]. LSNA technology typically samples in the time-domain such that Fourier transformation of the signal envelope allows the phase of individual mixing products to be referred to known reference tone, for example the lower fundamental tone of two-tone CW stimulus. LSNA technology in particular, and more generally, time-domain load-pull, forms an advanced element of load-pull practice offering deeper insights than available from standard load-pull measurements.

3.1.6 Dynamic Range Optimization

The dynamic range of a load-pull system for general power measurements, such as transducer gain, is established by the dynamic range of the power sensor at the source and load blocks. Only in exceptional circumstances is power sensor dynamic range a limiting factor in overall system dynamic range. In contrast, dynamic range under IM and ACPR characterization is bounded by the background noise of the signal analyzer and linearity of the reference PA or signal source, as illustrated in the plot of Figure 3.6. The region between these two boundaries establishes the dynamic range of IM and ACPR characterization.

In Figure 3.6, it is illustrated that, in general, the slope in the noise-limited region is approximately one-to-one while the slope in the linearity-limited region is approximately three-to-one. System dynamic range is optimized first by appropriate choice of power sensor and power capability. This is followed by optimizing signal conditioning with appropriate attenuation, coupling, and architecture.

3.2 The DC Power Source

Though seemingly innocuous, the DC power source is frequently responsible for substantial apparent signal impairment due to the presence of DUT low-frequency mixing products presented by the bias network [4]. Through interaction of bias network inductance and power source capacitance, a low frequency pole can be

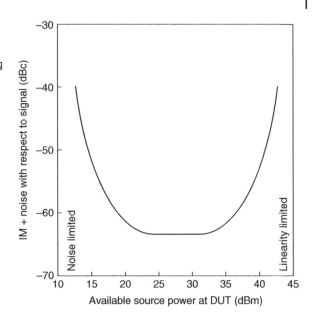

Figure 3.6 Typical dynamic range curve for IM and ACPR measurements illustrating noise-limited and linearity-limited boundaries. The region between these two boundaries establishes the dynamic range of IM and ACPR characterization.

introduced that is approximately equal to the envelope rate of the DUT stimulus modulation. Introduction of low-frequency mixing products near this pole exposes the (complex) envelope to rapid nonlinear phase changes, yielding asymmetry in IM and ACPR that is an artifact of the load-pull system, this impairment often referred to as bias network memory. Thus, it is quite probable that the load-pull system can mask the true linearity performance of the DUT, particularly when large capacitors are randomly placed in the bias network, presumably for decoupling or charge storage.

3.2.1 Charge Storage, Memory, and Video Bandwidth

The network between the DC power source and the DUT reference plane can be modeled as distributed resistance and inductance. To reduce the effects of these impairments, substantial capacitance is often placed near in the bias network, near the DUT, with the bias supplied over twisted pair or coaxial cable. The charge storage provided by the capacitance, to supply current under dynamic modulation, must be balanced against the introduction of memory into the bias network. The bandwidth at which this memory becomes significant is defined as the video bandwidth (VBW) of the bias network, and is usually defined[3] as the point where the phase asymmetry present under two-tone stimulus is 2°, as illustrated by Figure 3.7.

---

3 This definition is arbitrary, for example, in Section 2.2.3, we used 1 dB asymmetry in IM_3.

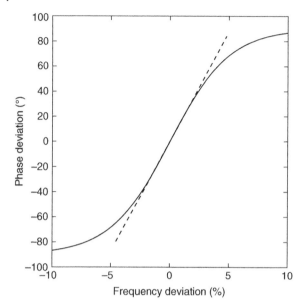

Figure 3.7 Response to a two-tone stimulus illustrating a common definition of video bandwidth (VBW) when the tone spacing yields 2° of asymmetry in the third-order mixing products. Note that the lower sideband leads in phase and the upper sideband lags in phase with respect to the linear phase response shown. It is this fact that produces intermodulation asymmetry, as described by Sevic and Steer [4].

There are two techniques available for characterization of the VBW of Figure 3.7. The first method characterizes the load-pull system itself, using measured s-parameters. In high power load-pull, in which the DUT internal pre-matching capacitance can interact with distributed bias network inductance, the *in situ* method from Noori is recommended, as it fully captures the effect of this interaction [8].

3.2.2 Load-Pull of True PAE

As a general rule, the passive load-pull system cannot measure true *PAE*, due to both source tuner heating loss and mismatch loss at the fixture and source tuner interface [9]. However, by performing an s-parameter measurement at an arbitrary reference-plane and resetting the reference-plane to the input of the DUT, large-signal input impedance can be measured with arbitrary precision, from which true *PAE* follows as

$$\eta_{PAE} = \eta \left(1 - \frac{1}{G_p} \right) \tag{3.5}$$

where η is the drain/collector efficiency of the DUT and G_p is the DUT power gain, defined as

$$G_p = \frac{G_T}{(1 - |\Gamma_{in}|^2)} \tag{3.6}$$

where $|\Gamma_{in}|$ is the large-signal DUT input impedance. In the absence of a vector receiver for large-signal input impedance measurement, true *PAE* can be approximated by substituting transducer gain for power gain in Eq. (3.6) as

$$\eta_{PAE} \approx \eta \left(1 - \frac{1}{G_T} \right) \tag{3.7}$$

which yields reasonable accuracy if the DUT exhibits transducer gain substantially larger than 10 dB or the DUT input impedance is reasonably close to 50 Ω. The errors are rigorously quantified in [9]. Finally, note that in contrast to passive load-pull, the active load-pull architecture will in general be able to measure true *PAE*, due to the presence of tuned receivers within its source block.

3.2.3 The Effect of DC Bias Network Loss

In a properly characterized load-pull system, the difference between actual DUT gain and measured gain can drive to an arbitrarily small value, within the resolution and repeatability of the power measurement. The effect of DC bias network losses, manifested as an instantaneous reduction in the bias voltage at the DUT reference plane, results in an apparent reduction in DUT power and gain.

The simplest method to reduce this error is simply an increase in the bias voltage by an amount approximately equal to the expected average voltage drop across the bias network. This works particularly well in CW load-pull. Load-pull with stimulus exhibiting substantial amplitude modulation requires a better estimate of the average bias current, usually by a true RMS ammeter with an integration time fast enough to track the modulation envelope. By monitoring this current, and knowing the DC resistance of the DC bias network, the effective bias voltage at the DUT reference-plane can be measured.

3.3 The ΔG_T Method of System Verification

Load-pull system performance verification is required to establish an acceptable level of accuracy and to assess long-term drift, each of which contribute to uncertainty in revealing actual DUT performance. Verification is usually composed of comparing system transducer gain with a known THRU standard, over the entire Smith chart, which is then compared to the THRU *s*-parameters, whose transducer gain is assumed to be the reference standard. Load-pull system transducer gain deviation from the THRU *s*-parameters can be related to dependent and derived parameters, usually PAE, providing a direct method of establishing how much uncertainty is acceptable for a particular application.

Comparing the transducer gain produced by the load-pull system to the transducer gain of a known standard is known as the ΔG_T method of load-pull system verification [1–3]. It is a widely adopted method that is accurate, repeatable, and simple, with sufficient generality for both passive and active load-pull.

The ΔG_T method compares the measured transducer gain provided by the load-pull system to that predicted by cascading the s-parameters of the each of the blocks of the load-pull system shown in Figure 3.2. Under the assumption that the transducer gain established by s-parameters is correct, any deviation from this transducer gain is defined as the ΔG_T, as illustrated by Figure 3.8, showing source and load tuners and a known, or defined, THRU. The ΔG_T is defined as

$$\Delta G_T = \frac{G_T|_{LP}}{G_T|_{VNA}} \tag{3.8}$$

where $G_T|_{LP}$ and $G_T|_{VNA}$ refer to the transducer gain provided by load-pull measurement and s-parameters, respectively, of the THRU of Figure 3.8.

The THRU standard must be known, or defined, for the ΔG_T method. If the test is conducted by concatenating the two fixture halves, it may be defined as an ideal THRU standard. If the THRU standard is inserted to emulate a DUT package, with a THRU standard in place of the packaged DUT, then the s-parameters of the THRU must be known. Similarly, in an on-wafer environment, the probe and substrate s-parameters must be known.

Note that a significant feature of the ΔG_T method is that all sources of potential s-parameter variation and interaction of each element of Figure 3.2 are captured and reduced to one number. Often times, a poor ΔG_T obtains only at extreme mismatch conditions, where small errors in de-embedding become relatively more significant; it is for this reason that thorough s-parameter characterization of each element of Figure 3.2 be conducted prior to assembly of the load-pull system, to identify and isolate potential sources of poor ΔG_T.

To demonstrate the ΔG_T method, consider a microstrip test-fixture with each fixture halve concatenated together to form an ideal THRU. In this case, the s-parameter matrix of the THRU of Figure 3.8 is

$$S = \begin{bmatrix} 0 & 1 \\ 1 & 0 \end{bmatrix} \tag{3.9}$$

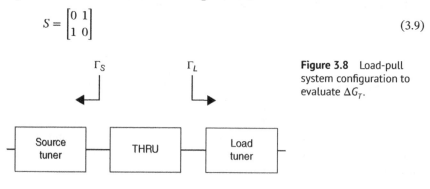

Figure 3.8 Load-pull system configuration to evaluate ΔG_T.

and ΔG_T is simply the mismatch loss between the source and load tuner

$$\Delta G_T = \frac{(1 - |\Gamma_S|^2)(1 - |\Gamma_L|^2)}{|1 - \Gamma_S\Gamma_L|^2} \tag{3.10}$$

Now source and load impedance may be swept over the entire gamma-domain and contours of ΔG_T plotted on the Smith chart, for example. To help isolate sources of error, if one or the other tuner of Figure 3.8 is set nominally to 50 Ω, then mismatch loss reduces to

$$\Delta G_T = 1 - |\Gamma|^2 \tag{3.11}$$

Under this condition, mismatch contours substantially displaced from 50 Ω indicate an element of the load-pull system that has become not well matched or whose actual s-parameters do not match the measured s-parameters. Other diagnostic tricks are possible.

In general, at higher mismatch factors, ΔG_T becomes significant. Similarly, source and load impedance may be fixed, say at an extremely high *VSWR*, and available source power swept, showing a response similar to Figure 3.9.

Table 3.1 illustrates a range of commonly acceptable ΔG_T values for different applications [1]. It is evident a major benefit of the ΔG_T method is quantitative treatment of systemic uncertainty of the load-pull process. By establishing a boundary on impedance variation against a known standard, a similar boundary of variation can be attached to measured transducer gain and PAE. The ΔG_T

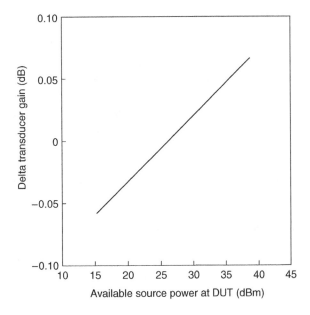

Figure 3.9 Typical ΔG_T response for a well-calibrated sub-1 Ω high-power 2 GHz load-pull system with quarter-wave pre-matching [10].

Table 3.1 Examples of ΔG_T and resultant $\Delta\eta$ for various applications [1].

Application	ΔG_T (dB)	$\Delta\eta$ (%)
Sub-1 Ω load-pull for high-power wireless load-pull	±0.10	±1.0
On-wafer active load-pull for 77 GHz model validation	±0.20	±1.5
In-fixture load-pull at $\approx$50 Ω	±0.50	±2.0

method provides the necessary ΔG_T for the acceptable variation of transducer gain and PAE, which depends on the user, the application, and other factors.

3.4 Electromechanical Tuner Calibration

All of the load-pull examples thus far have tacitly assumed a rectangular grid of impedance states for load-pull and source-pull, with no consideration given to an optimum configuration with respect to data acquisition time or reducing interpolation error. Optimizing the configuration of impedance state of load-pull is related to the traveling salesman problem, where it is required to visit a prescribed number of cities during a sales trip while minimizing some objective function, such as time spent traveling between cities or time money spent on airfare [11]. Application to load-pull adds an additional complication related to physical probe and carriage displacement, best described in cylindrical coordinates, versus the simplicity of a rectangular impedance state grid.

As a general rule, data acquisition time for large Smith chart regions is minimized by adopting a polar configuration of impedance state, as this reduces substantially an exchange of distinct probe displacement and carriage displacement operations, corresponding to tuner reflection-parameter magnitude and phase. An impedance state configuration in rectangular coordinates requires inordinate probe and carriage displacements to approximate rectangular coordinates, whereas in polar coordinates, the probe can be held approximately fixed for an each ring, constituting $\lambda/2$ around the Smith chart, progressing next to an adjustment of the probe and another complete $\lambda/2$ phase ring.

3.5 Closing Remarks

Load-pull system performance verification is required to establish an acceptable level of accuracy and to assess long-term drift, each of which contribute to uncertainty in revealing actual DUT performance. The ΔG_T method of load-pull system verification is based comparing system transducer gain with a known THRU standard, over the entire Smith chart, which is then compared to the THRU *s*-parameters, whose transducer gain is assumed to the reference standard. Load-pull system transducer gain deviation from the THRU *s*-parameters can be related to dependent and derived parameters, usually PAE, thereby providing a direct method of choosing how much uncertainty is acceptable for a particular application.

References

1 Noori, B. and Sevic, J.F. (2006). Load-pull: accuracy versus reality. IEEE Radio and Wireless Symposium.

2 Sevic, J.F. (2007). Theory of high-power load-pull characterization for RF and microwave transistors. Chapter 7. Boca Raton, FL: CRC Press, Inc.

3 Sevic, J.F. and Burger, K. (1996). Rigorous error analysis for low impedance load-pull, from Internal Notes at Qualcomm.

4 Sevic, J.F. and Steer, M.B. (1998). A novel envelope termination method for ACPR optimization of RF and microwave power amplifiers. IEEE Transactions on Microwave Theory and Techniques Symposium Digest.

5 Majerus, M. and Simpson, G.M. (1997). Input impedance measurement under large-signal load-pull conditions. Proceedings of the Automatic RF Techniques Group.

6 Radio corporation of America radio transmitter tube manual. p. 54, 1956.

7 Sevic, J.F. (2005). Phase measurement of IM for PA lineartiy optimization. IEEE MTT Micro-Apps Seminar.

8 Noori, B. (2004). A simple method of calculating VBW in-fixture. IEEE Radio and Wireless Symposium.

9 Sevic, J.F. (2005). Measuring true PAE with an automated load-pull system. *Application Note 5C-065. Tech. Rep.* Maury Microwave Corporation.

10 Sevic, J. F. (1996). A rigorous two-tier fixture calibration method for low impedance load-pull. Automatic RF Techniques Group Conference.

11 Applegate, D.L. and Bixby, R.E. (2006). *The Traveling Salesman Problem: A Computational Study*. Princeton, NJ: Princeton University Press.

4

Load-Pull Data Acquisition and Contour Generation

Systematic, efficient, and comprehensive data acquisition of transistor performance, versus terminating impedance, frequency, bias, power, and possibly other
parameters, is the central component of load-pull. By exposing to the transistor to
various combinations of fundamental source and load impedances, and various
harmonic terminations, patterns emerge that ultimately converge to unique
resolution of optimum performance and the associated impedances bestowing
this performance. Data acquisition establishes gradient information for optimal
performance trade-off, identifies maximum transistor performance limits, and
yields geometric regions in the Smith chart where performance requirements
simultaneously occur, the final product of load-pull being the source and load
impedances terminating the transistor at the fundamental and harmonics at an
associated bias.

Figure 4.1 illustrates the set of six fundamental parameters of load-pull data
acquisition. The set is complete when harmonic terminations and DC power are
added, which enables the transistor performance to be unambiguously characterized. Each of these are classified as independent, dependent, or derived. Independent parameters, which may be nested, are used to directly establish the operating
state of the transistor during load-pull. Dependent parameters are those parameters that are directly observable, or measurable, not used to establish the operating
state of the transistor but represent principal data for subsequent optimization and
matching network design. Derived parameters are the progeny of independent and
dependent data and may often be processed by instruments or software to yield a
particular performance measure.

For single-stage design, measurement of transistor input impedance, Z_{in} is
not required for matching network synthesis. However, for multistage design
using an interstage matching network, versus hybrid combining, measurement
of Z_{in} is necessary. Measurement of large-signal input impedance is rigorous
and physically meaningful, as well as being simple to measure [1]. The output
impedance of the transistor of Figure 4.1 is deliberately left undefined, as it is in

The Load-Pull Method of RF and Microwave Power Amplifier Design, First Edition. John F. Sevic.
© 2020 John Wiley & Sons, Inc. Published 2020 by John Wiley & Sons, Inc.

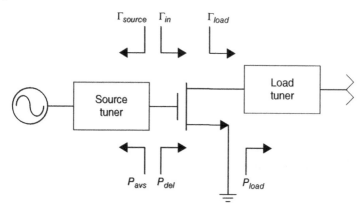

Figure 4.1 Canonical load-pull system defining independent variables from which all measurement data are derived. DC conditions at the drain (collector) and gate (base) form the basis of DC power for efficiency calculations. For the present load-pull method that transistor output impedance and its available power need not be known nor defined.

general unnecessary for load-pull method described presently. As a corollary, it is important to recognize that the output impedance of the transistor, assuming it can be defined rigorously, does not equal the conjugate of the load impedance.

The present chapter begins with treatment of load-pull with a single set of contours and then moving to sets of two more or contours. Swept, contingent search, constrained optimization, and parametric methods are next covered. Harmonic load-pull is presented as a special case of fundamental load-pull. The extremely powerful concept of logical-set operations on multivariable data for optimum geometric matching network design is also covered. Utilization of load-pull contour data is specifically treated in Chapter 5, as the primary tool for matching network design from load-pull data. Matching network design using load-pull data is described in Chapter 6.

4.1 Constant Source Power Load-Pull

Fixed available source power is the most popular method of load-pull, providing rapid convergence to a solution for simple applications, such as continuous wave (CW) modulation seeking maximum load power and power-added efficiency (PAE), while also providing a starting point for advanced data acquisition methods such as constrained optimization for ACPR–PAE load-pull. This method holds available source power constant while the independent parameters of Figure 4.1 are systematically varied, usually being fundamental source impedance, fundamental load impedance, and bias. Convergence to optimum source and load terminating impedances, and bias, is usually achieved in two or three iterations.

4.1.1 Load-Pull with a Single Set of Contours

While single-contour load-pull is conceptually simple, it nevertheless is extremely powerful, being capable of rapidly identifying maximum power, maximum gain, and maximum PAE, along with the associated load and source impedances delivering the associated performance. Maximum load power is established directly by the transistor load-line, so data acquisition generally begins with load-pull, followed by source-pull. To first-order, once the load impedance has been established for target load power, which is not necessarily the transistor's maximum peak envelope power (PEP) or CW power, gain is established by source-pull. Though load-pull is conducted at the fundamental and at harmonics, the process is in all but exceptional cases iterative, in that load-pull is followed by source-pull, with this loop repeating until convergence is achieved in resolving desired performance, subject to constraints, such as bandwidth versus impedance transformation ratio.

Fixed available source power load-pull requires initial estimates for source impedance, available source power, and quiescent current. Several options for their identification exists. As a general practice, it is recommended to extract the transistor s-parameters, when practical, before initiating load-pull. From s-parameters, a starting point for source impedance is identified by exposing the transistor to the Cripps analysis to identify a starting load impedance, Γ_{Cripps}. Using

$$\Gamma_{in} = S_{11} + \frac{S_{21}S_{12}\Gamma_{Cripps}}{1 - S_{22}\Gamma_{Cripps}} \tag{4.1}$$

leads to an estimate of Γ_{source}, where $\Gamma_{source} = \Gamma_{in}^*$ and Γ_{Cripps} is an initial estimate for load impedance from the Cripps analysis and source power is sufficiently backed off to assume not only linearity but to protect the transistor from damage during initial load-pull.

From the Cripps analysis and s-parameters, initial load-pull impedance states are established, shown in Figure 4.2. Transistor performance is measured at each distinct load impedance, with contours fitted to the region to the interior of the convex boundary comprised of the collection of impedance states. The dashed circle represents the maximum *VSWR* the tuner is capable of synthesizing, where two points in the grid are shown to exceed the maximum *VSWR* boundary, and are therefore excluded from measurement data for creating contours. Contour closure is a vital load-pull objective, as maximum and minima can be uniquely resolved. Nonclosure is remediated by test-fixture pre-matching or adopting internal tuner pre-matching. Transistor scaling is not recommended because the distributed-network present on the transverse axis of the transistor creates

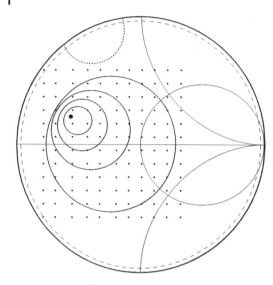

Figure 4.2 Load power contours illustrating the maximum power impedance, contour closure, and the impedance-state grid for data acquisition. The large dashed concentric circle at the edge of the Smith chart boundary is the maximum *VSWR* the tuner can develop while the partial dashed circle near the top of the Smith chart is the load stability circle at the load-pull frequency.

interference phenomena that leads to inconclusive results in both scaling of power and load impedance.

Unstable regions over frequency can be isolated from the s-parameters to limit the impedance states of Figure 4.2 to regions of unconditional stability, as the load stability circle illustrates. Absent an estimate based on the Cripps method for input impedance or s-parameter data, a reasonable starting point, for consistency, is to adopt a source impedance of 50 Ω. In following the theme of searching for measurement consistency, the measured S_{21} can be compared to load-pull G_T at 50 Ω as further system verification.

Identification of optimum quiescent current is crucial for optimum performance as it directly establishes the operating class of the power amplifier (PA). In the vast majority of applications, quiescent current of $0.1I_{dss}$ is a reasonable starting point, though, ultimately, swept-bias load-pull must be employed to identify the quiescent current that provides not only optimum gain but especially optimum linearity. Swept-bias load-pull is treated later in the present chapter.

Figure 4.3 illustrates the iterative process of load-pull and source-pull to identify optimum power, gain, and the associated load and source impedances at which these occur for a transistor representative of 1 W at 3.4 V. Each iteration serves to illustrate not only the several steps of the load-pull process but also issues that typically arise, such as impedance states not entirely capturing maxima or minima.

The first iteration, on the top row, illustrates load power contours from the initial impedance states estimated from Γ_{Cripps}. The source impedance is set to Γ_{source}, illustrated by the square in the source-domain; Γ_{Cripps} is also illustrated as a square in the load-domain. The initial load impedance estimate provided

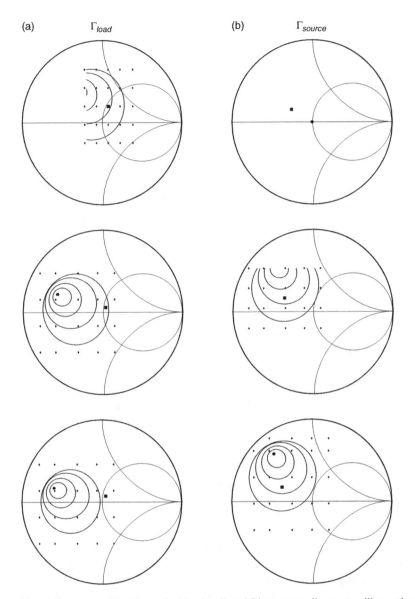

Figure 4.3 Three iterations of (a) load-pull and (b) source-pull contours illustrating convergence to maximum power and gain. The first iteration is based on Cripps estimate, shown as a square, also illustrating the impedance states did not resolve power and gain maxima, thus exhibiting open contours. Subsequent expansion of the impedance states for both load-pull and source-pull enabled power and gain maxima to be uniquely resolved.

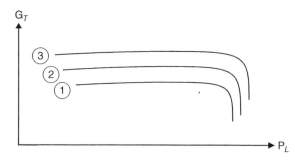

Figure 4.4 Transducer gain versus load power for each of the three load-pull iterations of Figure 4.3. As convergence is achieved for optimum load and source impedance, both maximum power and gain increase.

reasonable coverage for generating a number of contours, yet open contours preclude resolving maximum power and the impedance at which it occurs. The resultant transducer gain versus load power trajectory for this first iteration is shown in Figure 4.4 as trajectory one.

The second iteration, shown in the second row of Figure 4.3, evolves from the first by moving the load-pull impedance-state domain to the left to capture the maximum power impedance, by resolving closed contours, followed by source-pull. The results show that the load-domain impedance states have correctly been chosen to generate closed contours, thereby resolving maximum power and the associated impedance at which it occurs. Note that for fixed available source power load-pull, the load power and transducer gain contours are coincident. The source-pull contours illustrate a reasonable estimate from Γ_{source}, calculated by Eq. (4.1), though it is evident, at least one additional iteration is necessary to generate closed contours. The resultant transducer gain versus load power trajectory for the second iteration is shown in Figure 4.4 as trajectory two, where it is seen that the maximum power load impedance has resulted in an increase in small-signal, large-signal gain, and maximum load power.

The third iteration is shown in the third row of Figure 4.3, illustrating a slight change the optimum load impedance maxima, which is usual, since the reverse isolation of a microwave transistor induces input–output coupling.[1] As well, adequate impedance state coverage has been achieved in the source-domain for closed source-pull contours for closed contours, thus resolving maximum gain and its associated source impedance. The resultant transducer gain versus load power trajectory for the third, and final, iteration is shown in Figure 4.4 as trajectory three, where it is seen that simultaneous maximum power load and source impedance result in further additional increase in small-signal, large-signal gain, and maximum load power over the second iteration.

A solution for maximum power and gain was resolved in three iterations, which is typical for CW applications, such as GSM. Load-pull for linear applications,

---

1 Gate–drain (base–collector) capacitance and source (emitter) inductance are the two most common sources of finite reverse isolation.

such as WCDMA and LTE, can take many more iterations, and usually requires sophisticated data acquisition methods. Nevertheless, from the solution, maximum power and maximum gain, and their associated load and source impedances, have been identified. Maximum power or gain may not necessarily be the goal, however, leading to impedance selection along a load or source contour of constant load power. Because there exist an infinite number of impedance states along this contour, there exists an infinite number of matching network implementations delivering the load power described by this contour, leading to ambiguity in identifying a unique load impedance. Several approaches exist to identify a unique load impedance state, subject to additional constraints, most often matching network Q, as it relates to implementing a required impedance transformation ratio and bandwidth.

Fixed available source power load-pull, while conceptually simple, suffers from an inability to easily visualize transistor performance under constant compression or constant linearity. Moreover, holding available source power constant essentially increases by one the number of variables to be optimized in resolving a multi-parametric solution involving, for example, load power, gain, PAE, and ACPR (adjacent channel power ratio) simultaneously. Superimposing multiple contours, next discussed, addresses the limitations of single-parameter load-pull by enabling simultaneous visualization and optimization of multiparameter constraints, some mutually exclusive, such as load power and PAE. The present treatment has assumed constant available source power and fixed frequency. The indispensable topic of source impedance trajectory and load impedance trajectory identification versus frequency is presented in Chapter 5, where it will be combined with the present methods to yield the required impedance trajectory data for matching network synthesis and physical implementation.

4.1.2 Load-Pull with Two or More Sets of Contours

Load-pull and source-pull with multiple coincident contour sets superimposed on the Smith chart provides both a substantial improvement in simultaneous optimization of the performance of several variables while also laying the foundation of several advanced methods of data acquisition. Whereas load-pull with a single contour set provides a unique parametric optimum, and its associated source and load impedance, two or more sets of coincident contours graphically and analytically permit solutions providing an optimum spanning several variables simultaneously over terminating impedance, frequency, bias, and power.

Load-pull with multiple sets of contours commences identically to load-pull with a single set of contours, including identification of initial load and source impedances using the Cripps method, or equivalent, selection of bias current, and creation of a seed set of source and load impedance states. The presence of

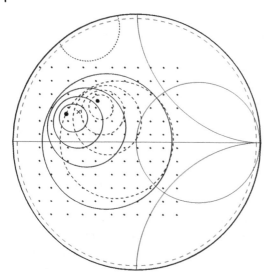

Figure 4.5 Load power (solid contours) and PAE (dashed contours) contours illustrating the maximum power and PAE impedances, contour closure, and the impedance-state grid for data acquisition. The large dashed concentric circle at the edge of the Smith chart boundary is the maximum *VSWR* the tuner can develop while the partial dashed circle near the top of the Smith chart is the load stability circle at the load-pull frequency. The X symbol illustrates a possible trade-off point for maximum PAE for a given power, at a fixed frequency.

multiple sets of contours tends to inflate the impedance-state domain associated with a single set of contours, so additional consideration is necessary to capture regions of interest for multiple parameters. Load-pull with multiple sets of contours is particularly useful when optimizing signal quality against PAE, each of which are mutually exclusive constraints. This approach is often done in parallel with swept bias current optimization and envelope load-pull to simultaneously optimize each independent parameter the ACPR–PAE trade-off condition critical linear RF PA designs.

Figure 4.5 illustrates load power contours of Figure 4.2 superimposed with PAE contours, a combination useful for CW applications to identify the so-called trade-off point representing, for a given transistor size, maximum PAE for a given load power. The impedance states were selected based on a Cripps analysis and *s*-parameters measurements to establish the stability circle shown. The impedance-state domain is considerably larger than load-pull with a single set of contours, especially when exploring load power and PAE.

Figure 4.6 extends the iterative process illustrated by Figure 4.3 to include PAE. The first iteration, on the top row, illustrates load power contours and PAE contours. Identification of initial impedance states for load-pull and source-pull with multiple sets of contours combines the Cripps method for the load power estimate and estimates the expected location of the remaining contour sets found from physical principles, such as maximum PAE being on a steeper load-line than that for maximum load power. The initial load impedance estimate provides reasonable coverage for generating a number of both load power and PAE contours, yet the open contours preclude resolving maximum power and maximum PAE and the impedances at which they occur.

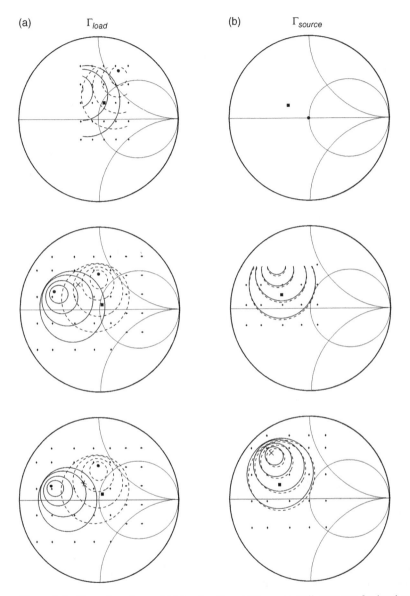

Figure 4.6 Three iterations of (a) load-pull and (b) source-pull contours for load power and PAE illustrating convergence to maximum power, gain, and PAE. The first iteration is based on Cripps estimate, shown as a square, also illustrating the impedance states did not resolve power and gain maxima, thus exhibiting open contours. Subsequent expansion of the impedance states for both load-pull and source-pull enabled power, gain, and PAE maxima to be uniquely resolved. The X symbol illustrates a possible trade-off point for maximum PAE for a given power, at a fixed frequency.

The second iteration, shown in the second row of Figure 4.6, evolves from the first by moving the load-pull impedance-state domain to the left to capture the maximum power impedance and to the upper right to capture the maximum PAE impedance. The impedance-state domain of the source-pull has also been expanded, where it is noted that load power and gain circles are very nearly coincident, as is to be expected for fixed available source power load-pull. The load-pull results illustrate that multiple contour sets enable parametric trade-off optimization, substantially compounding the power of multiple contour data over virtually all other methods of RF PA design. In identifying the impedance state where multiple criteria are simultaneously Pareto optimum, for example illustrated by the X shown in the load-domain, maximum PAE for given load power is resolved, as is the associated load impedance at which it occurs.

The third iteration is shown in the third row of Figure 4.6, illustrating sufficient impedance-state coverage for identification of optimum performance in both the source-domain and load-domain. Several significant observations follow. A distinct Pareto optima for load power and PAE is now evident from this iteration, shown by the X symbol, illustrating necessary compromise in choosing a given PAE, and it associated load power, and the impedance delivering this performance, which could be further constrained to matching network Q. Since movement away from this impedance state, in any direction, results in a decrease of load power and PAE, this impedance state has the extremely powerful property that it represents a local simultaneous maxima for load power and PAE, and is defined as the trade-off point for load power and PAE. This property will be further generalized in Chapter 5 by deriving the impedance trajectory between maximum load power and PAE, which will be defined as the optimum impedance trajectory. It is important to observe this trajectory is not the shortest line between the two optima, but is rather the trajectory that is everywhere orthogonal to each contour set.

Because transducer gain and load power contours are coincident for fixed available source power, the optimum source impedance requires trade-off only in essentially one variable. Commonly defined as the trade-off state, it represents generally the maximum PAE for a given transistor size at an associated maximum load-power. Section 4.1.3 will illustrate the versatility of multiple contour sets in the source-domain when signal quality is explored.

Fixed available source power load-pull with multiple sets of contours offers a substantial improvement in the ability to resolve Pareto optima of several parameters simultaneously. In seeking a Pareto optimum trade-off point, an impedance state is selected so that its gradient is everywhere negative, in contrast to displacement along one or more contours where it may remain constant. By removing the condition that available source power be held constant, this method can be easily extended to optimization or three or more parameters simultaneously, most usually being PAE, gain, signal quality, and frequency.

4.1.3 Load-Pull for Signal Quality Optimization

Simultaneous signal quality, load power, and PAE load-pull, with fixed available source power, provides a highly visual analytical and graphical method for optimum impedance trajectory identification that is virtually unrivaled by other RF PA design methods, while also laying the foundation of the advanced methods discussed later. Superposition of signal quality, load power, and PAE contour identifies on the Smith chart a unique point, or region, where all three criteria can simultaneously meet, or exceed, target requirements. Signal quality broadly encompasses many definitions, including IM distortion, ACPR and ACLR, static AM–PM, and error vector magnitude (EVM). It is possible even to extend load-pull to dynamic AM–PM for load-pull coupled with digital pre-distortion (DPD) characterization.

Signal-quality load-pull begins in a manner identical to the previous methods, including identification of an initial seed set of source and load impedance states. A general trend with signal-quality load-pull is load-line establishment to support the PEP[2] of the signal followed by, usually, extensive source-pull to identify the optimum trade-off between gain and linearity. Since the optimum linearity source impedance state is substantially displaced from optimum large-signal gain of CW load-pull, it is not uncommon for source-pull to be done over the entire Smith chart as an initial estimate. Identification of optimum bias current is a key component of signal-quality load-pull and will be discussed Section 4.4.2. For the present treatment, it is assumed that a preliminary optimum bias current has been identified.

Figure 4.7 illustrates load-pull contours of signal quality, load power, and PAE. The signal quality contours are deliberately left open to illustrate they may not exhibit circular or elliptical shapes, and in fact may not in general close over the load-pull or source-pull domain. For the present example, the signal quality contours are configured to reflect a minima, meaning that to the interior of the convex hull of the signal quality contours points to a minimum, for example the smallest ACPR in dBc. A general trend with signal quality is that it tends to improve in the vicinity of maximum power and degrade in the vicinity of maximum PAE, this being a consequence of maximum power reflecting maximum symmetrical swing, hence maximum PEP.

Figure 4.8 extends the iterative process to include signal quality, which for simplicity is represented as one generic contour set shown by the dotted contour. While the Cripps method can continue to be used to seed the load-domain, no simplified analytical method is available for the source-domain for signal-quality source-pull. To address the need to quickly identify optimum source conditions, it

---

2 To first order, PEP is approximated by sum of the required average power and peak-to-average ratio (PAR) of the modulated signal.

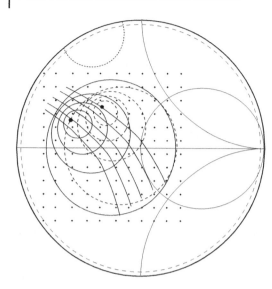

Figure 4.7 Load power (solid contours), PAE (dashed contours), and ACPR (solid lines) contours illustrating the maximum power and PAE impedances, ACPR contours, contour closure, and the impedance-state grid for data acquisition. The large dashed concentric circle at the edge of the Smith chart boundary is the maximum *VSWR* the tuner can develop while the partial dashed circle near the top of the Smith chart is the load stability circle at the load-pull frequency.

is common practice to simply seed the entire Smith chart for the source-domain, with a relatively low-state density, to aid in identifying optima and their associated impedances. It is seen from the example that near complete coverage is obtained, revealing signal quality optima that is substantially displaced from transducer gain and load power maxima. Note that signal quality optima is used, versus maxima or minima, in keeping with the generic definition of signal quality.

The second, and final, iteration expands significantly the load-domain, yielding adequate coverage for both source-pull and load-pull. Graphical identification of a load impedance state where the gradient of three contour sets is simultaneously zero – a stationary point – poses somewhat of a challenge, so the simplification often made is to hold signal quality constant and from that contour identify an impedance state where load power and PAE are stationary.

Consider now the load power and PAE trade-off state denoted by the X in the load-domain of the second row of Figure 4.8, representing a maximum PAE for a given load power. Convergence to this state represents a stationary point for maximum PAE and load power. The signal quality contour intersecting X is approximately orthogonal to the stationary point denoted by X. Orthogonal contours, or a contour orthogonal to a trajectory composed of stationary points, exhibit the highly useful property that the power–PAE stationary point can be displaced along the signal quality contour, holding its value constant, while simultaneously locating the optimum power–PAE trade-off impedance state, providing decoupled optimization.

The optimum impedance state of the second iteration of the source-domain illustrates a scenario where an optimum signal quality contour is now tangent to

(a) Γ_{load} (b) Γ_{source}

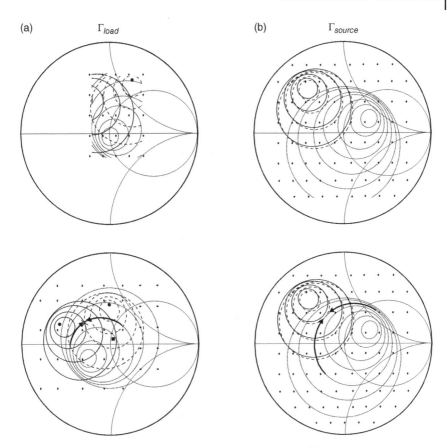

Figure 4.8 Two iterations of (a) load-pull and (b) source-pull contours for load power (solid contours), PAE (dashed contours), and signal quality (dotted contours) illustrating convergence to optimum simultaneous power, gain, PAE, and signal quality. The initial source-pull covers the entire Smith chart to quickly identify trends in optimum signal quality. Note from the first iteration that maximum transducer gain and maximum signal quality, e.g. minimum ACPR, are substantially displaced.

the pair of contours representing, in this example, load power and transducer gain, versus being orthogonal. A similar process is followed, though in this case the load power and gain will substantially vary during displacement along a constant signal quality contour. Optimum performance in this example is obtained by identifying a set of load power and gain contours that are tangent to the target signal quality contour. Note that the present example specifically illustrated signal-quality contours orthogonal to load-pull power contours and tangent to source-pull gain contours as two extreme cases; in general, signal quality contours will be neither strictly orthogonal nor strictly tangent to power, gain, or PAE contours.

4.1.4 Large-Signal Input Impedance

Designing a multistage RF PA with the load-pull method requires a slight modification to the single-stage design process, from which identification of the source and load impedances terminating the transistor at the fundamental, and possibly harmonics, are sufficient. Since source impedance uniquely establishes performance, through its interaction with the transistor input port, the single-stage method simply requires an input matching network to match from 50 Ω to the source impedance specified by load-pull, with no consideration to the input impedance of the transistor.

Such conditions do not, however, exist in the multistage RF PA, since the interstage matching network is not terminated by 50 Ω, having instead to match from the input impedance of the upstream stage, as illustrated in Figure 4.9, to the load impedance of the downstream stage. The situation is further complicated since there is a tendency to assume that the transistor input impedance is simply the complex conjugate of the source impedance identified by load-pull. Not only would this violate the assumption of weak nonlinearity, but because conjugate match corresponds to maximum power transfer, and maximum power transfer is seldom the goal of source-pull, assuming complex conjugate would lead to impaired performance, especially linearity.

Thus, a source impedance identified by source-pull as optimum for some ensemble of conditions, for example maximum simultaneous PAE, gain, and linearity, does not infer, or even suggest, that it can be assumed that the input impedance of the transistor is simply the conjugate of the source impedance identified by load-pull. Identification of the transistor input impedance must therefore be obtained through alternative rigorous means.

The configuration due to Majerus provides such a systematic and rigorous empirical method of resolving large-signal transistor input impedance under realistic available source power, load power, bias, and load impedance conditions [1]. The method associates with each measurement, whether source-pull, load-pull, or swept parameter, a large-signal input impedance measured at an arbitrary

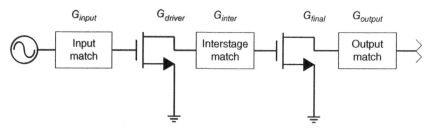

Figure 4.9 Generic multistage PA line-up illustrating interstage matching network and relevant impedance definitions.

reference-plane that subsequently completes the set of all required parameters for interstage matching network design. The load-pull method of RF PA design for multistage applications requires access to the large-signal transistor input impedance, at an appropriate reference-plane, to enable design of the interstage matching network.

4.2 Fixed-Parametric Load-Pull

Fixed available source power load-pull is an elementary, yet powerful, method belonging to the broader class of fixed-parametric load-pull distinguished by holding one, or more, parameters – independent, dependent, derived – to explore device under test (DUT) performance under the most general conditions short of swept available source power. It is especially useful in those applications where available source power and load power must be decoupled from one another, such as identification of constant gain compression and constant ACPR. It is also useful in complexity reduction by holding an arbitrary parameter fixed, usually a target parameter such as load power, as the number of contour sets comprising the optimization space is reduced by one, simplifying graphical synthesis of the optimum impedance trajectory.

Fixed-parametric load-pull is implemented by specifying a target condition, and associated tolerance range, for the fixed-parameter variable, and then iterating available source power to achieve the target condition[3] until the fixed-parameter falls within the target range. Iteration occurs at each impedance state to emulate a virtual fixed-parameter environment. Fixed load power, fixed gain compression, fixed signal quality, and fixed peak–average ratio are treated presently, though the procedure is sufficiently general to be applied to any parameter that can be measured.

In addition to varying available source power as the iteration control variable, synthetic fixed-parametric load-pull can be implemented as a post-processing operation by slicing a multidimensional data set composed of load-pull, source-pull, and swept available source power. This method, while the most general, requires an extended investment of measurement time. It is especially useful for process development, over multiple wafer sites and runs.

4.2.1 Fixed Load Power

The process of load-pull under fixed load power is similar to fixed available source power yet substantially more effective and graphically lucid, since optimization

---

3 There is no reason why another variable could not be iterated, such as drain voltage for polar PA applications. Nevertheless, available source power is the most common, and is the variable used in this book unless otherwise stated.

of the remaining variables occurs at the target load power. Once a region of the Smith chart has been identified supporting the target load power, the general process is explore this region exclusively to locate an optimum stationary point for the remaining variables, usually PAE and signal quality. Fixed load power provides the option to graphically assess the robustness of various DUT sizes and technologies through the relative size of the region enclosing the target load power. A relatively larger area will tend to provide more resilience to mismatch and process, voltage, and temperature (PVT) variation, whereas a relatively smaller area represents lower die cost at the risk of lower yield.

Because fixed load power iterates available source power, several conditions exist that can lead to corrupted data when the requested available source power to produce the target load power is not achievable. To mitigate, or eliminate, each of these impairments, it is recommended first to confirm that the DUT is capable of delivering the target load power along the boundary of the region being explored. It should also be verified the DUT is not being overly compressed. The most common impairment is the driver PA of the load-pull system, not the DUT, may be overly compressed or simply incapable of producing the necessary power to overcome the available loss of the source tuner. The last condition is especially relevant in signal quality characterization as the driver PA can impair the apparent signal quality presented by the DUT. Existence of any one of these conditions will artificially suppress the target load power and skew the resultant contours.

Figure 4.10 illustrates regions where added attention should be placed to avoid. The annular ring represents the region where the tuner available loss increases rapidly, as illustrated in Figure 2.15. When synthesizing an impedance in this

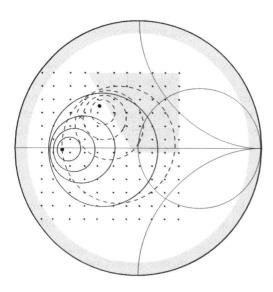

Figure 4.10 Illustration of regions requiring special attention during fixed load power from another, with the gray annular ring representative of high available loss and the second gray region the location of relatively high impedances that can induce premature saturation and subsequently lower power capability.

annular region, the power capability of the driver PA can rise substantially due to a rapid increase in available tuner loss.[4] The additional gray region represents the location of relatively higher impedance with respect to optimum, hence reducing the maximum power capability of the DUT. Unless it is intended to explore regions of maximum PAE, this region should be carefully considered. While in general the regions are not necessarily precisely located as illustrated by Figure 4.10, the essential concern is to be aware of the sources of an inability to reach the target load power and plan the load-pull strategy accordingly.

4.2.2 Fixed Gain Compression

Because fixed available source power and fixed load power are not decoupled, or independent of each other, both methods are incapable of uniquely resolving fixed gain compression, a common metric found in applications spanning mobile GSM to DPD for base-stations. Fixed gain compression identifies contours of constant gain compression, such as P_1 and P_{sat} based on iterating both available source power and load impedance, which is a for proxy load power capability. Through these variables, contours of constant gain are identified. Transducer gain is almost always used, though there is no reason power gain cannot be used.

Identification of optima, including stationary points, proceeds in a manner identical to the methods previously discussed. The most common combination of stationary points for fixed gain compression is load power and PAE, yielding required load power and PAE at the target gain compression. Like load-pull with fixed load power, attention must be given to ensure the DUT is capable of producing the target gain compression and falls within the range of the minimum required load power and PAE.

4.2.3 Fixed Peak–Average Ratio

Many PA applications rely on signal conditioning algorithms to reduce the peak–average ratio of the signal being amplified, thereby enabling operation closer to saturation and subsequently higher PAE. Fixing the peak–average ratio renders it a virtual independent variable that allows regions of maximum PAE and load power to be identified, based on the target peak–average ratio. Implementing fixed peak–average ratio requires a power meter or receiver capable of measuring peak–average ratio, as well as a signal source capable of synthesizing the signal with the appropriate resolution of signals with statistically infrequent peak powers.

The pseudo-stochastic nature of peak–average characterization yields an ensemble of multiple, statistically identical,[5] instantaneous signal trajectories exhibiting

---

4 This property is another advantage of pre-matching.
5 To the second moment, meaning the autocorrelation function is invariant over the ensemble.

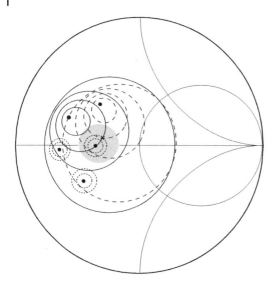

Figure 4.11 Contours of load power, PAE, and peak–average ratio. Note the phenomena of multiple maxima, illustrating the need for fixed peak–average ratio load-pull to be preceded by fixed available source power load-pull to assist in identification of an appropriate load power and PAE stationary point to perform fixed peak–average ratio load-pull. The circular region in gray is identified as the region for fixed peak–average ratio since it encloses the load power and PAE stationary point and while providing a locally unique solution.

the same target peak–average ratio, as defined on the complimentary cumulative distribution function (CCDF). This can introduce an ambiguity in the form of multiple maxima on the Smith chart, inhibiting the search for a unique peak–average ratio that is simultaneously on a stationary point of load power and PAE. Ambiguity is easily resolved by performing first a fixed available source power load-pull, as illustrated in Figure 4.11, to graphically isolate a unique peak–average ratio maxima in proximity to the target load power and PAE stationary point, defined by the gray circle. Fixed peak–average ratio load-pull is subsequently confined to this region, with the process repeated for source-pull.

A final consideration for peak–average load-pull is the adoption of a stationary PN sequence that excites all possible modes of the DUT, essentially a PN sequence long enough to contain an ensemble of instantaneous signal trajectories that statistically replicate the stochastic nature of the actual application environment. The stationary property ensures that the ensemble's instantaneous trajectories leading up to a peak power event remains invariant, and the DUT is thus exposed to repeatable and consistent stimulus to assure repeatability as the measurement progresses from one impedance state to another. The consequence of nonstationary is maxima that appearing to drift over the Smith chart, rendering convergence impossible to a unique solution.

4.2.4 Fixed Signal Quality

Fixed signal quality load-pull is broad colloquial term that includes harmonic distortion, intermodulation distortion, ACPR/ACLR, EVM, code-domain power (CDP), and AM–PM. Fixed signal quality most often is reserved as the final step of, possible several, different load-pull procedures that include swept power and swept bias. Swept power is utilized to identify the optimum die size (active area) of the transistor for the target PAE and signal quality. Swept bias identifies the appropriate bias current to yield optimum signal quality, usually recognized by distinct interference that occurs near compression [2–4].

To begin fixed signal quality analysis, near-optimum regions of both the source-domain and load-domain are first isolated using swept bias and swept power. Because linearity is strong function of source impedance, it is recommended to start with fixed signal quality source-pull, followed by fixed signal quality load-pull, and proceed in an iterative fashion until convergence is achieved.

4.2.5 Treating Multiple Contour Intersections

The treatment thus far has intentionally provided examples where stationary points are unambiguously defined by the tangency of two distinct contours, for example load power and PAE. In practice, such well-ordering seldom occurs, though such tangency is in fact representative of necessary and sufficient conditions for simultaneous optimality. To illustrate, consider Figure 4.12 illustrating the target PAE contour intersecting target load power contour at two locations, denoted by X. This graphically illustrates that the DUT provides more load power than necessary at the target PAE. The obvious conclusion is that the DUT power capability can be scaled back – i.e. reduce its physical area – so that the target load power contour is precisely tangent to the target PAE contour, being a stationary point for the target specifications. Alternatively, this condition can be interpreted that additional load power is available at simultaneously higher PAE. Though

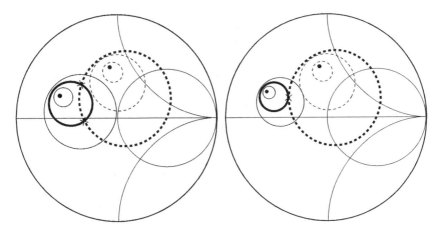

Figure 4.12   Illustration of multiple contour intersections and their interpretation.

this state initially appears desirable, recall that the target load power for many applications implicitly assumes not only minimum but also a maximum load power, typically as a proxy to bound die cost. Hence, the first conclusion tends to be correct interpretation.

The second Smith chart illustrates the situation with die size chosen to provide precisely the target load power at the target PAE, resulting in the unique single-valued stationary point shown by the X. As would expected, the load power contours have contracted somewhat and the real component of the maximum load power impedance state has increased, as would be expected for a reduction in DUT power capability. The result moves beyond Pareto optimum to global optimum in that the DUT is now sized for target load power and target PAE at the lowest possible die area, and hence cost, for a given transistor technology.

An extension of this method is to add die cost as an optimization variable. Since die cost is directly proportional to die area, formulation of the problem in these terms would add another dimension to the optimization problem to control for die cost.

4.3 Harmonic Load-Pull

A comprehensive theoretical and empirical framework of the optimum harmonic termination has been built around the load-pull method of RF PA design [3–5]. From this rigorous, yet nonmathematical framework, it is possible for nearly cost-free and substantial improvements in PAE and linearity by identifying the optimum harmonic terminating impedance at the load, in particular, followed by the source. Aligning with the general theme of the book – management versus analysis of nonlinearity – elementary theory is used to locate a general area on the Smith chart for load-pull exploration, followed by extensive region-specific load-pull to identify the optimum performance and the impedances bestowing this performance.

The primary value of load-pull in balancing management versus of analysis of nonlinearity is its singular ability to unambiguously and systematically resolve optimum harmonic terminations while simultaneously reconciling effects such as relative phase of the transistor harmonics, parasitics internal to the transistor, and its package, and external distributed effects. These effects work *in toto* inducing a deviation in practice, from the ideal, of the harmonic phase delivering optimum PAE and linearity. Virtually all analyses of PAE improvement make no distinction between the relative phases of the harmonics with respect to its fundamental. Because of this, particularly at RF and microwave frequencies where transit time is a large fraction of the period of the carrier, load-pull is ideal for empirical identification of the optimum harmonic termination. Thus, there may be a phase deviation from ideal even if the reference-plane is set directly at the intrinsic active area of the transistor. The presence of internal matching, package delay effects, and parasitics external to the transistor will also move the harmonic phase from ideal.

This section treats harmonic load-pull for PAE and linearity enhancement. Second harmonic load-pull comes first, followed by third-order and simultaneous second-order and third-order harmonic load-pull. Though seldom necessary, higher-order load-pull is briefly treated. The highly important topic of base-band load-pull is also treated. Because even-order mixing products have a base-band component, empirical optimization of this impedance is necessary in applications that extend video bandwidth (VBW), typically associated with DPD wrapped around a linear RF PA.

4.3.1 Second Harmonic Load-Pull

Harmonic load-pull and source-pull almost always occur following optimization at the fundamental, and like fundamental load-pull, requires iteration for convergence. Because the Fourier components of a signal are coupled in the sense that pre-optimum second harmonic impedance termination may induce a small change in the optimum fundamental impedance termination, some iteration will be in general be necessary, especially in attempting to identify conditions for ultra-high PAE over 80%. The process of harmonic load-pull can follow any of the optimization methods described in the present chapter, though it usually starts with fixed available source power and progressing to fixed load power. Final evaluation with harmonic load-pull is best done using fixed load-power to minimize its influence on resolving maximum PAE.

Second harmonic load-pull for high PAE targets a high-reflectivity termination with respect to the fundamental impedance, though the efficacy of harmonic termination depends more on the relative ratio of impedances at the fundamental and harmonics than the absolute value of the harmonic termination impedance. This implies that infinite *VSWR*, as provided, for example, by active load-pull, is not necessarily required, and in any event, this mismatch cannot be replicated physically. Second harmonic load-pull for optimum linearity will generally target a high-reflectivity termination, though impedance states to the interior of the Smith chart can yield optimum linearity and should not be excluded from the exploration space. Figure 4.13 illustrates an impedance state distribution useful for second harmonic load-pull optimization. The states shown are often simply chosen as the maximum reliable *VSWR* the tuner is capable of producing.

Because the primary independent variable is second harmonic phase, performance visualization is best done in rectangular coordinate with dependent and derived parameters plotted versus second harmonic phase with load power fixed.

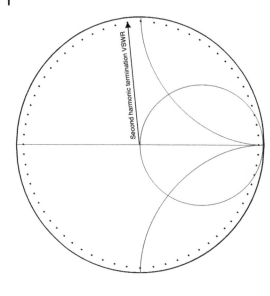

Figure 4.13 Common impedance state distribution for harmonic load-pull exploring optimum PAE. A usual distribution would be 36 or 72 states, yielding a phase step of 10° or 5°, respectively. The *VSWR* for harmonic load-pull will usually be the maximum *VSWR* the tuner is capable of producing, as shown by the vector.

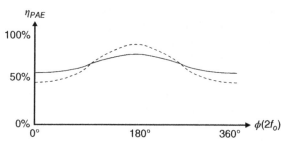

Figure 4.14 *PAE* versus second harmonic phase for fixed available source power (dashed line) and fixed load power (solid line).

Figure 4.14 shows rectangular coordinates PAE versus second harmonic phase for both fixed available source power and fixed load power with the fundamental impedance termination set for maximum PAE. For this general example, the peak PAE occurs at a slight offset from that predicted by theory due to an intrinsic second harmonic phase offset, showing the utility of harmonic load-pull in rapidly resolving the optimum second harmonic termination. Figure 4.14 also illustrates that under fixed load-power load-pull that the peak–peak PAE spread contracts, compared to fixed available source power.

Second harmonic load-pull for signal-quality optimization will generally explore the entire Smith chart, to ensure there are no optima in the interior. As with PAE optimization, it is often best to hold load power fixed, or, alternatively, hold signal quality fixed and search for maximum power and PAE, as would be done when identifying also the optimum transistor size. Source-pull is performed in a similar manner, though performance improvement will be more modest.

4.3.2 Third-Harmonic Load-Pull

For the vast majority of applications, especially those where $f_o \approx 0.1 f_t$, second harmonic load-pull provides virtually all PAE improvement possible, thus rendering third-harmonic load-pull an optional second-order improvement. In those applications where $f_o \ll 0.1 f_t$ or where maximum PAE is absolutely necessary, then third-harmonic load-pull is warranted. It may also be warranted for reference or benchmarking purposes for process development or process development kit (PDK) development. Advanced gallium nitride (GaN) semiconductor technology is an important example of the necessity of third-harmonic load-pull, as the literature has illustrated [6].

Third-harmonic load-pull proceeds in an identical fashion to second harmonic load-pull, after optimum fundamental and second harmonic terminations have been identified. Because interaction exists between the harmonic components, essentially cross-modulation of each Fourier component, some iteration between fundamental and harmonic impedance optima is necessary. With time-domain capability and load-pull, it is recommended to visualize the influence of harmonic tuning and its effect on suppression or accentuation and the flattening of the voltage (or current[6]) waveform.

4.3.3 Higher-Order Effects and Inter-harmonic Coupling

Load-pull at harmonics higher than third-order is only necessary in exceptional applications, particularly for frequency-domain synthesis of the Class E PA due to Raab [9]. Present state-of-the-art is passive load-pull limited to fourth-harmonic, this being a practical limitation versus physical or mathematical limitation.

Inter-harmonic coupling is the process whereby the conditions at one harmonic alter the conditions at other harmonics by cross-modulating the Fourier components of the load-pull stimulus. While the effect is generally small, in those applications requiring maximum performance, it implies that iteration between load-pull at each harmonic is mandatory.

4.3.4 Baseband Load-Pull for Video Bandwidth Optimization

Base-band influence on in-band linearity of a microwave PA was demonstrated by Mass, with further development by Sevic and Steer [3–5]. The optimum design of the bias network, which establishes the base-band impedance, remains a central issue for successful linear PA design. The presence of even-order nonlinearities

---

6 This book has provided a simplified treatment of Class F and other related topologies that rely on harmonic terminations. There exist so-called inverse topologies that operate on current instead of voltage [7, 8].

in odd-order mixing products causes in-band linearity to be influenced by out-of-band impedance terminations at baseband, at approximately half of the maximum instantaneous modulation frequency, and at the harmonics. While Volterra analysis can be invoked to provide an estimate of the optimum baseband impedance, it is generally faster and more accurate to perform baseband load-pull. The alternative is to simply maximize VBW using linear-phase filter theory for bias network design, which is presented Chapter 6.

The vector IM diagram of Figure 4.15 illustrates a quantitative framework to understand the influence of the bias network on VBW, using a two-tone stimulus. Baseband load-pull involves adjusting the effective length and phase of the even-order vector nonlinear term, comprising one component of three odd-order mixing products, by altering the magnitude and phase of the admittance at

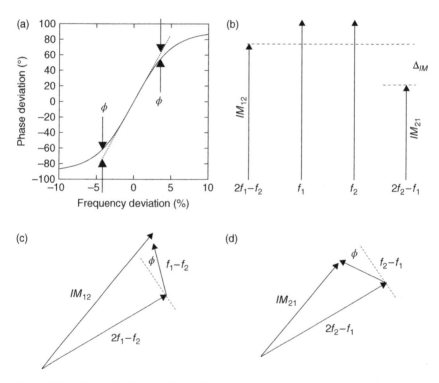

Figure 4.15 Vector IM diagram illustrating how the baseband impedance influences in-band linearity. Maximizing video bandwidth pushes out any resonances of the bias network that expands the length of the second-order IM vector, whereas optimization using baseband load-pull systematically seeks out an impedance antiparallel to the vector sum of the other two mixing product vectors. (a) Phase response deviation, (b) asymmetric intermodulation products, (c) upper IM vector diagram, and (d) lower IM vector diagram.

baseband. Because the optimum magnitude and phase is difficult to establish *a priori*, exploration of the entire Smith chart is recommended, since optimum linearity obtains by making the even-order vector antiparallel to the remaining two vector terms. Base-band load-pull is performed using stimulus identical to that used in the end application or product and is usually done simultaneously with second harmonic load-pull as they are both, ultimately, a manifestation of the same underlying nonlinearity.

4.4 Swept Load-Pull

Swept load-pull represents the most general class of load-pull data acquisition, capable of identifying any simultaneous combination of optima with full decoupling between available source power, load power, bias, and frequency. It is the foundation of advanced optimization methods that slice multidimensional data structures where several variables are simultaneously held fixed, as well as geometrical methods based on logical operators between contour sets. Various classes of load-pull can be nested, for example swept power and swept bias, to identify optimum linearity and PAE trade-off conditions. Swept load-pull is frequently automated, to enable unattended data collection, that might otherwise involve many hours of acquisition and processing.

4.4.1 Swept Available Source Power

Swept available source power load-pull provides deep insight into DUT behavior, the weakly nonlinear region to deep compression, at various terminating impedance states and bias conditions. When coupled with swept frequency and harmonic load-pull, it provides the most general picture of DUT performance and the terminating impedances bestowing that performance. In addition to providing a multitude of data, including constant compression, constant gain, and constant signal quality, swept power load-pull is useful in analyzing advanced architectures, such as the Doherty PA, where swept power and PAE are provided versus load impedance to assist in optimum Doherty network synthesis.

Swept available source power begins similar to fixed source power load-pull, including defining an initial load-pull domain and source-pull domain. Supplementing the establishment of initial conditions is specification of the available source power range. It is recommended to check a few load impedance states near the boundaries, particularly near the maximum PAE state, as this will produce substantial gain compression. Gain compression, with its associated higher voltage swing at gate and drain, can initiate various transistor breakdown mechanisms, leading to DUT failure.

4.4.2 Swept Bias

Swept bias can vary both voltage and current, though drain (collector) current is the most common due to its direct effect on gain and signal quality [3, 5]. Current is swept indirectly by varying the gate voltage or base current, depending on the transistor technology. Swept drain (collector) voltage is often used for Doherty, EER, envelope tracking (ET), and envelope following (EF) analysis to evaluate DUT performance versus voltage, particularly gain and power linearity an AM–PM.

Load-pull with swept bias is usually preceded by fixed power load-pull to identify a region where a preliminary target load-power and PAE stationary point has been first located. With this target region, bias current is swept at each source and load impedance state for fixed load power to identify an optimum IM/ACPR/ACLR bias current.

4.4.3 Swept Frequency

Identification of the source and load terminating impedances over frequency is the ultimate goal of load-pull. Following identification of the optimum bias condition and optimum source and load impedances, at a nominal frequency, swept frequency load-pull enables unique resolution of the source and load impedance trajectories bestowing target performance. It is these trajectories that are replicated by the matching network that ultimately replicate target performance.

4.5 Advanced Techniques of Data Acquisition

This section combines builds on the previous methods into two extremely powerful methods of simultaneous optimization. The first method is well-suited for simultaneous unconstrained optimization of several variables by creating irregular geometric regions of the Smith chart where target criteria are simultaneously true based on simple logical set operations, such as AND, OR, and NOT. The second method is specifically intended for process development that must consider not only performance optimization of a single transistor but also possibly tens or hundreds spanning multiple wafer sites, multiple wafers, and multiple epitaxial stacks. This method essentially introduces a swept-index variable denoting physical location to resolve trends in subtle process changes either in primarily material growth or photolithography. This advanced topic of data management over many runs of load-pull is used also to illustrate state-of-the-art in what is impossible when the power of systematic load-pull data acquisition is combined with the power of multidimensional numerical optimization methods and search routines.

4.5.1 Simplified Geometric-Logical Search

Geometric-logical search performs common logical operations on contour sets to illustrate where in the Smith chart the resultant logical expression is true or, if so defined, false. The result is generally an irregularly shaped region in the Smith chart that, when combined with the circle-mapping property of the bilinear transform upon which the Smith chart is based, leads to an efficient and highly visual method of matching network design. The most common logical set operations are AND, OR, NOT, while occasionally XOR is useful. The method is especially useful for rapid linear RF PA design once an optimum bias current has been identified using swept bias. Two approaches are presented. The first approach is useful for identifying approximate regions where multiple criteria are satisfied, and is particularly useful for gauging relative transistor sizes and ease-of-matching. The second method, treated in Section 4.5.2, is considerably more powerful by combining load-pull data from multiple frequencies to enable a synthetic impedance trajectory versus frequency that supports multiple target criteria, enabling arbitrary attainment of flatness over frequency.

To start the procedure, it is first required to attach with each contour a target range, such as minimum load power, minimum PAE, and maximum ACPR. It helps to bound each parameter, otherwise an unclosed region may result, even if one of the boundaries is essentially a dummy value. A search algorithm then seeks out in the gamma-load and gamma-source domains the region where each criteria simultaneously satisfy the logical expression, the shape being defined as the convex hull of the impedance states defining the boundary of the region. In many cases, this can be done visually, though most commercial load-pull packages have this basic function included as a standard feature.

Figure 4.16 illustrating load contours of PAE and ACPR for fixed load power with the shaded irregular region representing the boundary where ACPR and PAE are simultaneously satisfied. For fixed frequency, a matching network presenting a load impedance anywhere in this region will deliver performance within the specified boundaries. The relative size of the region, for fixed load power spanning transistors of varying sizes, is an excellent proxy for optimizing cost versus performance. Nevertheless, flatness over the region may be unacceptable, which is resolved by the synthetic geometric-logical search next.

4.5.2 Synthetic Geometric-Logical Search

Synthetic geometric-logical search expands on the simplified method by combining load-pull data from multiple frequencies into one search region create a synthetic impedance trajectory versus frequency that simultaneously satisfies the parametric boundaries established by the target specifications. The

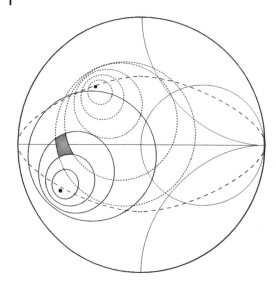

Figure 4.16 Illustration of geometric-logical search method showing the boundary where PAE and ACPR constraints are satisfied simultaneously for fixed load power and fixed frequency. The elliptical dashed lines are contours of constant Q, used later.

resultant impedance trajectory represents conditions that when replicated by the physical matching network yields performance targets over frequency. This method is often done near the conclusion of the transistor evaluation process to converge on the final source, load, and interstage matching network responses over frequency.

The synthetic geometric-logical search begins by establishing the target boundaries for each parameter, based on minimum or maximum targets plus an additional delta. It is important to define the band of each parameter to be small as an initial value to aid in achieving the flatness target over frequency. This step is repeated at each frequency over the band of interest, and at harmonics, if desired, with each region then simultaneously superimposed on the Smith chart to synthesize a truth region versus frequency delineating the region where all parametric criteria are satisfied. This result represents the impedance trajectory to be replicated by the physical matching network.

A useful property of bilinear transformation, which relates impedance and reflection coefficient on the Smith chart, maps circles from one domain to the other domain. For example, a 2 : 1 *VSWR* mismatch circle at the PA reference-plane appears as a 2 : 1 mismatch circle at the final-stage transistor reference-plane, normalized to a different reference impedance. This property, coupled with synthetic geometric-logical search, enables a graphical method of assessing bandwidth and mismatch effects for a given matching network order as illustrated by (Figure 4.17).

Figure 4.17 Illustration of the synthetic geometric-logical search method showing three regions where PAE and ACPR constraints are satisfied, representing low-band, mid-band, and high-band. The tolerance band for each parameter must generally be smaller than the geometric-logical search method to approximate a line versus region. A blow-up is shown in the circle to illustrate this, along with an approximate impedance trajectory.

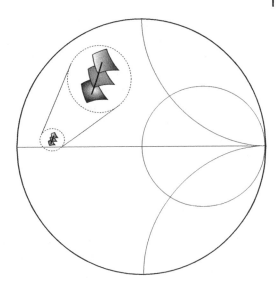

4.5.3 Multidimensional Load-Pull and Data Slicing

The data acquisition methods previously treated involve a maximum of two independent variables.[7] One independent variable is dedicated to sweeping source or load impedance, with the second independent variable dedicated usually to available source power, to yield contour data for dependent and derived data. Multidimensional load-pull generalizes load-pull with two independent variables by providing a systematic framework for data acquisition and visualization, based on arbitrary nesting of independent variables, resulting in a comprehensive data-structure representing DUT performance spanning all possible combinations of impedance states and stimuli. A significant benefit of multidimensional load-pull is the ability to extract a constant-parametric cross-section spanning the data-structure, similar to fixed-parametric load-pull, but without the requirement of iterating available source-power to hold dependent and derived data fixed to a target value. Multidimensional load-pull can be faster, when a complete understanding of DUT performance is necessary, and is also useful for operator-free fully automated data acquisition. Due to its extremely high speed, active load-pull is ideal for multidimensional load-pull.

Multidimensional load-pull adopts a systematic structure, defined by a sweep-plan specifying each swept parameter, its associated sweep boundaries and step-size, and its nesting-level within the sweep hierarchy. Fundamental load-pull

---

7 There is no reason for this other than being an artifact of the historical evolution of modern commercial load-pull.

and source-pull are typically placed at the bottom of the sweep-plan hierarchy, because of the relatively long dwell time associated with physical movement of the tuner; this ensures physical tuner physical movements are minimized, minimizing overall data acquisition time. Working up from the bottom of the sweep-plan hierarchy, simultaneous swept source and load impedance are followed by swept available source power, swept bias, and swept frequency, at each combination of source and load impedance. Harmonic load-pull and source-pull are usually performed subsequent to multidimensional fundamental load-pull, as a distinct operation, since the coupling of the fundamental swept parameters have only weak influence on the general shape and optima of parameters strongly influenced by harmonic impedance, such as PAE.

Data-slicing is the process of partitioning the data-structure over an arbitrary set of cross-sections, holding each cross-section constant to a target constraint values. A simple example is shown in Figure 4.18, illustrating a three-dimensional data-structure composed of swept available source power, swept frequency, and swept fundamental load reflection coefficient. The slice parallel to the frequency-gamma plane, illustrated by the gray surface, represents load-pull at constant available source power, P'_{avs}, spanning frequencies $[f_1\ f_2\ f_3\ f_4]$, and load reflection coefficients $[\Gamma_1\ \Gamma_2\ \Gamma_3\ \Gamma_4]$. For each frequency, this data structure is equivalent to fixed available source power load-pull. Similarly, the surface shown in Figure 4.19 illustrates load-pull at fixed frequency f' with swept available source power and swept fundamental load reflection coefficient corresponding to swept load-pull.

Multidimensional load-pull is applied to two broad classes of data acquisition. The most common class is fully general DUT characterization over bias, frequency, and power to isolate optimum performance, particularly the linearity-efficiency trade-off for constant load-power. The sweep-plan below illustrates a common

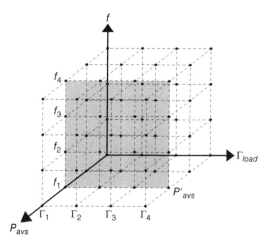

Figure 4.18 Illustration of data-structure composed of three independent dimensions to illustrate a data slice for fixed available source power.

Figure 4.19 Illustration of data-structure composed of three independent dimensions to illustrate a data slice for fixed frequency.

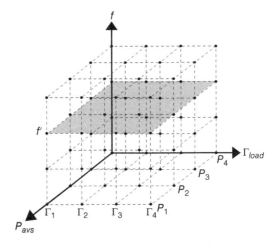

hierarchy for implementing multidimensional load-pull for identifying optimum DUT performance. Data slicing is applied to isolate optimum performance, and from there, the techniques of the present chapter are applied to isolate the optimum source and load impedance trajectories.

- Swept frequency
- Swept bias
- Swept available source power
- Swept fundamental source and load impedance

The second class of data acquisition addressed by multidimensional load-pull is cataloging wafer-map performance over various fabrication lots, wafers, and wafer locations to isolate optimum fabrication parameters, such as an heterojunction bipolar transistor (HBT) epi-stack delivering maximum simultaneous PAE and ACLR. The sweep-plan below illustrates a general hierarchy for implementing multidimensional load-pull for wafer-mapping. Data slicing is used to identify the optimum lot, wafer, or wafer location yielding optimum performance.

- Wafer or DUT identifier (e.g. lot number, x–y location, etc.)
- Ambient temperature
- Swept bias
- Swept frequency
- Swept available source power
- Swept fundamental source and load impedance

4.5.4 Min–Max Peak Searching

Global peak searching is a data acquisition method to rapidly identify a single parametric maxima or minima and the impedance bestowing this performance.

The underlying algorithms employed by peak search employ data interpolation to synthesize a gradient vector pointing to optimum performance conditions, and the associated impedance, bypassing the data-intensive data acquisition methods of previous sections. This substantially reduces the time necessary for identifying optimum DUT performance, though it does not provide necessary detail for establishing optimum trade-off over several parameters simultaneously. Peak searching is ideal for quick assessment of DUT performance and to identify a region on the Smith chart to initiate a load-pull session.

Successful min–max peak search requires a robust interpolation algorithm to provide well-conditioned data for the gradient estimation. Since the gradient is a directional derivative, approximated by finite differences, it is susceptible to data acquisition noise and numerical noise that under certain circumstances yields a vector that does not point to the correct min–max location. Two-dimensional interpolation that avoids rapid changes in the approximation of the surface of load-pull data over the Smith chart are generally preferred, particularly Laplacian interpolation and splines, each of which exhibit continuous first- and second-order derivatives.

4.6 Closing Remarks

Systematic, efficient, and comprehensive data acquisition of transistor performance, versus terminating impedance, frequency, bias, power, and possibly other parameters, is the central component of load-pull. By exposing to the transistor to various combinations of fundamental source and load impedances, and various harmonic terminations, patterns emerge that ultimately converge to unique resolution of optimum performance and the associated impedances bestowing this performance.

This chapter has introduced both basic and advanced methods of load-pull data acquisition, starting with a single set of contours and then moving to sets of two more or contours. Swept, contingent search, constrained optimization, and parametric methods were introduced to identify an optimum impedance that satisfies several criteria simultaneously. Harmonic load-pull was presented as a special case of fundamental load-pull. Utilization of load-pull contour data is specifically treated in the following chapter, as the primary tool for matching network design from load-pull data. Matching network design using load-pull data is described in Chapter 6.

References

1 Majerus, M. and Simpson, G.M. (1997). Input impedance measurement under large-signal load-pull conditions. Proceedings of the Automatic RF Techniques Group.

2 Sevic, J.F. and Steer, M.B. (1995). Analysis of MESFET spectral regrowth with a pi/4-QPSK modulated source. Proceedings of the IEEE International Microwave Symposium.

3 Maas, S.A. (1988). *Analysis of Nonlinear Microwave Circuits*. London: Artech House.

4 Stephen, B.N., Maas, A., and Tait, D.L. (1992). Intermodulation in hetero-junction bipolar transistors. IEEE Transactions on Microwave Theory and Techniques.

5 Sevic, J.F. and Steer, M.B. (1998). A novel envelope termination method for ACPR optimization of RF and microwave power amplifiers. IEEE Transactions on Microwave Theory and Techniques Symposium Digest.

6 Sheppard, S., Pribble, B., Smith, R.P. et al. (2006). High-efficiency amplifiers using AlGaN/GaN HEMTs on SiC. CS MANTECH Conference.

7 Cripps, S.C. (2002). *RF Power Amplifiers for Wireless Communications*, 2e. London: Artech House.

8 Cripps, S.C. (2008). *Advanced Techniques in RF Power Amplifier Design*. London: Artech House.

9 Raab, F.H. (2001). Class-E, Class-C, and Class-F power amplifiers based upon a finite number of harmonics. *IEEE Transactions on Microwave Theory and Techniques* 49 (8): 1462–1468.

5

Optimum Impedance Identification

Through systematic data acquisition under specific stimulus, automated load-pull yields sets of contours parametrized to impedance, frequency, bias, and power, rigorously describing complex nonlinear transistor behavior. Because stimulus and transistor embedding are realistic to arbitrary precision, each contour set represents a faithful embodiment of transistor behavior as a well-defined function of source and load impedance. Multiple contour sets plotted on the Smith chart provide a highly visual medium to identify explicit contour intersections, or regions of overlap, where underlying parameters are simultaneously optimum, possibly subject to additional constraints, such as bandwidth.

Explicit contour intersections, or regions of contour overlap, lead naturally to the concept of optimum impedance, which is defined as the impedance state that meets, or exceeds, all parametric objectives simultaneously. This is the impedance state that, when physically replicated by the matching network embedding the transistor, yields the optimal behavior identified by load-pull, net of matching network loss. The optimum impedance state is thus the key link between load-pull and optimal RF power amplifier (PA) design.

The present chapter graphically defines the optimum impedance state, and its generalization, the optimum impedance trajectory. Several key properties are presented, followed by an introduction to construction methods using load-pull contours. Graphical methods can be used for optimum impedance synthesis usually up to two parameters, possibly against a constant third parameter, such as average load power.

5.1 Physical Interpretation of the Optimum Impedance

In Chapter 4, the tangent point of a pair of contours, of two distinct parameters, was seen to represent a local optimum. The heuristic treatment was motivated by the desirable property that for a specified value represented by one contour, the

The Load-Pull Method of RF and Microwave Power Amplifier Design, First Edition. John F. Sevic.
© 2020 John Wiley & Sons, Inc. Published 2020 by John Wiley & Sons, Inc.

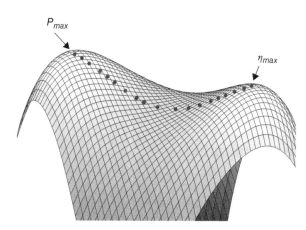

Figure 5.1 Three dimensional rendering of a CW optimum impedance trajectory illustrating the line that is simultaneously maximum for each parameter. Note that any displacement off the optimum trajectory, represented by the sequence of impedance states on the surface, except in its optimal direction, yields a decrease in at least one of the parameters.

tangent contour was simultaneously optimum. Qualitatively, this locally optimum is that point where incremental displacement from it yields a decrease in at least one of the parameters.

Figure 5.1 illustrates a 3D physical rendering of a continuous wave (CW) load-pull surface representing load power and power-added efficiency (PAE). The two peaks correspond to maximum load power and maximum PAE, respectively. Each of the discrete impedance states between the maxima, depicted as the dots, represent the optimum trajectory, since displacement in any direction not on the trajectory results in one of the parameters being suboptimal. This trajectory is defined as the optimum impedance trajectory and represents the well-known concept of trade-off impedance between maximum load power and maximum PAE.

Each impedance state on the optimum impedance trajectory yields the maximum PAE for a given load power. Similarly, for a given PAE, each impedance state yields an associated maximum load power, hence providing optimal trade-off between each parameter. Consistent with the heuristic definition of an optimum impedance trajectory, observe that displacement off the optimum impedance trajectory states may in general yield a decrease in both load power and PAE, depending on the relative displacement away from the trajectory. This illustrates the desired physical property of optimizing each parameter simultaneously. Note, however, that additional constraints will in general complicate the shape of an optimum impedance trajectory.

5.2 The Optimum Impedance Trajectory

The Smith chart is the most useful impedance-plane to depict the optimum impedance trajectory of Figure 5.1, since several parameters and constant-Q circles can be superimposed on it simultaneously, as well as facilitating the matching network synthesis methods described in Chapter 6. In this context, both the optimum impedance, and its generalization, the optimum impedance trajectory, are the key link between load-pull data and RF PA design. Figure 5.2 depicts a generic optimum impedance trajectory, composed of a set of optimum impedance states. The optimum impedance trajectory is constructed on the principle of orthogonality, with the trajectory everywhere orthogonal to the line tangent to the two contours composing the optimum impedance state. We begin in Section 5.2.1 by describing three fundamental properties of the optimum impedance trajectory.

5.2.1 Optimality Condition

Optimality is the state that exists when each parameter is simultaneously optimal with the other parameters. Though usually meaning maximum, optimal does not always mean maximum, however, as ACLR is optimum when minimized,

Figure 5.2 Graphical definition of the optimum impedance trajectory, illustrating its definition as the line everywhere orthogonal to the tangent points of adjacent contours connecting the two associated optima. The optimum impedance trajectory is the line composed of the set of optimum impedance states between the two contours, in the present example being load power and PAE.

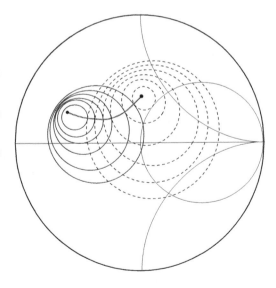

for example.[1] In the common situation of load power and PAE trade-off, optimality would be defined as maximum simultaneous load power and PAE. The orthogonality property produces optimality since the optimum impedance trajectory passes through the set of all tangency points that, collectively, produce optimum performance. The orthogonality property can also be applied to PAE and signal quality, for example, with load power held constant. This would yield an optimum impedance trajectory describing maximum PAE and associated minimum ACLR, at constant average load power.

5.2.2 Uniqueness Condition

Convexity may be invoked to prove that the optimum impedance trajectory is the one, and only, line that is everywhere orthogonal to adjacent contour tangency points, while terminating at the geometric center of each corresponding circle. This yields necessary and sufficient conditions required for treatment of graphical optimal impedance trajectory synthesis techniques in Section 5.2.3. As a corollary of the uniqueness condition and orthogonality property, the optimum impedance trajectory is in general curvilinear, which directly follows from the distinct, and mutually exclusive, physical properties of maximum load power, maximum PAE, and maximum linearity.[2]

5.2.3 Terminating Impedance

Independent and dependent parameters often appear as approximately circular, or elliptical, contours on the Smith chart.[3] For a pair of contours, the optimum impedance trajectory begins and ends on a respective contour's maximum, or minimum, thereby establishing its terminating impedances. Parameters represented by closed contours exhibit the useful property of uniquely distinguishing a maximum or minimum, illustrating the importance of closed contours.

An optimum impedance trajectory based on derived parameters, particularly those expressed as a function of two more independent or dependent parameters, or parameters representing signal quality, often appear as quasi-planar contours on the Smith chart. Figure 4.7 illustrates load power and PAE contours,

---

1 Signal quality for commercial wireless applications represents an excellent example of optimum implying just good enough for specification compliance, as there is generally no reward for exceeding a specification, especially if it sacrifices battery life or cost.
2 Maximum load power implies maximum symmetrical voltage and current swing; maximum PAE implies maximum compression of voltage and current; maximum linearity requires instantaneous support of peak envelope power (PEP). All of these conditions, for a current-mode RF PA, are mutually exclusive, and hence yield distinctly different contour sets.
3 Recall the Cripps analysis.

superimposed with adjacent channel power ratio (ACPR) contours illustrating there is no clear maximum or minimum on the interior of the Smith chart. It is possible that the quasi-planar ACPR surface is locally orthogonal to PAE contours, dramatically simplifying the optimization process, since ACPR and PAE are decoupled. This desirable phenomena does not in practice, however, always occur.

5.3 Graphical Extraction of the Optimum Impedance

Graphical extraction of the optimum impedance state exploits the geometric interpretation of that point where two parametric contours are mutually tangent. By extension, the optimum impedance trajectory is the set of all optimum impedance states, with the trajectory terminated by the respective optimal values for each parameter, everywhere orthogonal to the contours. To implement the previous heuristic treatment in a systematic and rigorous fashion, a, graphical technique is presently introduced that enables the optimal impedance state, and optimal impedance trajectory, to each be extracted from measured load-pull contours.

5.3.1 Optimum Impedance State Extraction

Figure 5.3 illustrates an abbreviated pair of contours, each with a distinct mutual tangent point. Though only two impedance states are shown, in general there are an infinite number of states, the selection of which is based on target performance. To locate the optimal impedance state graphically, each point where a mutual

Figure 5.3 Illustration of graphical optimum impedance state extraction, showing the tangent line at the point where two parametric contours are mutually tangent.

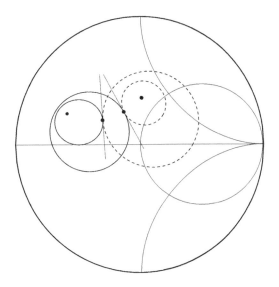

tangent obtains is selected as a target optimal state, as shown by the two distinct tangent lines. Because there are an infinite number of states, a constant must be added to locate the tangent point on a specific contour pair.

In this context, one parameter establishes the location along one contour, for example average load power, while the line tangent to the impedance state fixes the position of the second contour, establishing its associated value. Figure 5.3 illustrates tangent lines at the two tangent points. If the desired targets are not met simultaneously, then iteration is necessary, for example selection of a different contour or a change in emitter area, for example. Following this systematic, yet simple, graphical process enables repeatable and modestly accurate results. To motivate the graphical technique, consider now an example.

5.3.2 Optimum Impedance Trajectory Extraction

There are two techniques available for optimum impedance trajectory extraction. The first technique is an extension of Section 5.3.1, which traces an orthogonal trajectory to the mutual tangent points between each optima. This method can be done by hand, and reasonably accurate for many applications.

The second method, which is considerably more precise and systematic, plots in rectangular coordinates a pair of load-pull data parameters against each other, parametrized to impedance state. Consider Figure 5.4 illustrating load

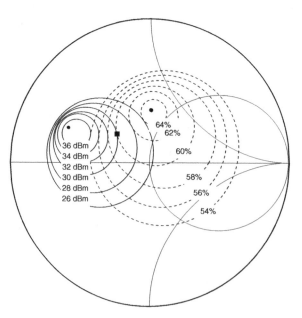

Figure 5.4 Load-pull contours for a 10 W GaN device, at 1.9 GHz, illustrating load power and PAE.

Figure 5.5
Rectangular-coordinate plot of average load power versus PAE. The underlying impedance states attached to the convex hull of the data represent of the optimum impedance trajectory between maximum load power and maximum PAE, with each state representing an optimum point that is simultaneously maximum load power and maximum PAE.

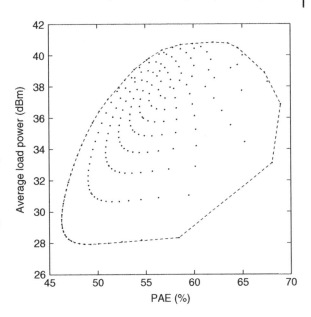

power and PAE of a 10 W Cree gallium nitride (GaN) high electron mobility transistor (HEMT).[4] Now consider Figure 5.5 plotting the load power and PAE in rectangular coordinates. In this illustration, each load impedance state is represented by a dot with the envelope enclosing all of the impedance states defined as the convex hull. The convex hull is a geometric construct defined such that a line intersecting two points on the hull do not enter the region on its exterior.

The optimum impedance trajectory obtains from associating the impedance points between the maxima of the two parameters, in the present example being average load power and PAE, to a trajectory in the gamma-domain. Note that the trajectory obeys the properties outlined previously, including terminating on maxima, simultaneous optimality of load power and PAE, and uniqueness of the trajectory, which follows the impedance states defined by the segment of the convex hull connecting the two maxima.

Figure 5.6 illustrates the optimal impedance trajectory of the 10 W GaN HEMT described by the load-pull contours of Figure 5.4. Note that this trajectory indeed describes the optimal trade-off in load power and PAE, is everywhere orthogonal to each contour, and terminates at maximum load power and PAE.

---

4 The author thanks Dr. Ray Pengelly for providing the measured GaN data.

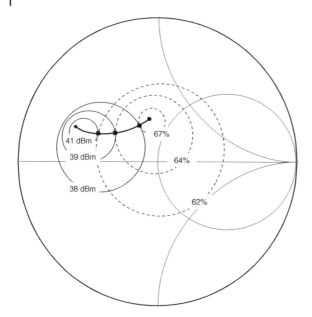

Figure 5.6 The optimal impedance trajectory of the 10 W GaN HEMT described by the load-pull contours of Figure 5.4. Note that this trajectory indeed describes the optimal trade-off in load power and PAE, is everywhere orthogonal to each contour, and terminates at maximum load power and PAE.

5.3.3 Treatment of Orthogonal Contours

First introduced in Figure 4.8, orthogonal contours are the polar opposite of tangent contours and play an important role in optimum impedance extraction. Orthogonal contours, or almost-orthogonal contours, represent the important case of parametric decoupling, in which one parameter is held constant, or almost constant, while another parameter is varied. This enables one parameter to be optimized while the second parameter is held essentially fixed.

Figure 5.7 illustrates the concept of orthogonal contours at the intersection of 28 dBm and 58% PAE. Any of the four trajectories illustrated by the arrow are approximately constant while the orthogonal contour varies. For example, consider displacement on the 28 dBm contour in which PAE varies from approximately 56–60%, illustrating that for a constant load power of 28 dBm, the optimum impedance state would likely be at the intersection of the 28 dBm contour and 60%. Similarly, for constant PAE of 58%, load power varies from 30 to 26 dBm, with the intersection of the 30 dBm and the 58% being the optimum impedance.

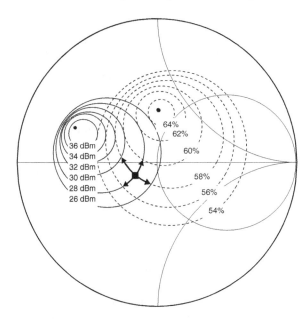

Figure 5.7 Illustration of the concept of orthogonal contours where one parameter is held approximately fixed and the other is allowed to vary.

5.4 Optimum Impedance Extraction from Load-Pull Contours

The groundwork for optimum impedance extraction has been laid, with the concepts of optimum impedance state and optimum impedance trajectory. Using these general concepts enables systematic and accurate graphical impedance extraction methods that simultaneously optimize average load power, PAE, signal quality, and other arbitrary parameters, under various operating conditions. The present section will apply these two concepts to various load-pull scenarios to enable optimization of up to three distinct variables.[5]

Recall the earlier definition of the three load-pull data classes: independent, dependent, and derived. Up to the present section, the presentation has implicitly assumed constant available source power, which is an independent variable. However, in many load-pull applications, it is desired to have a dependent variable held constant, thus making it a virtual independent variable. For example, to reduce simultaneous load-pull of load-power, PAE, and signal quality to

---

5 There is no limit to the number of variables that can be simultaneously optimized, but with four or more variables, the graphical method becomes tedious. Nonlinear optimization methods, as such Marquardt–Levenberg, can be employed for more sophisticated applications.

simultaneous load-pull of PAE and signal quality, load-power – a dependent variable – can be held constant. Similarly, it may be desired to hold gain compression of peak-to-average ratio (PAR) constant. This type of load-pull is implemented by varying available source power, at each impedance state, to hold the desired dependent variable constant, making it a virtual independent variable.

5.4.1 Simultaneous Average Load Power and PAE

Impedance identification of simultaneous maximum load power and PAE represents the most basic, yet highly useful, form of load-pull. Available source power is held constant, and source and load impedance are each iterated until convergence is achieved. Consider Figure 5.8 showing average load power and PAE for a final-stage GSM-900 mobile application. It is desired to achieve 36 dBm at 75% PAE with greater than 15 dB of transducer gain, G_T. There are two ways to identify the optimum impedance, both producing the same result. Consider first construction of the optimum impedance trajectory, which is illustrated by the curvilinear trajectory terminating at the two maxima. In this simple example, the optimum impedance state is at the tangent point of 36 dBm and 75% PAE.

An alternative method is to identify the contour that satisfies either of the two required targets. For example, starting with the 36 dBm contour, displacement along it until it is tangent to the target PAE requirement yields as well the desired optimum impedance. These two methods are based on the optimum impedance trajectory and the optimum impedance state method, respectively. In general, the contours may not be as well defined, though both methods should yield approximately the same impedance, in the present example being $\Gamma_{load} \approx 0.4\angle160°$.

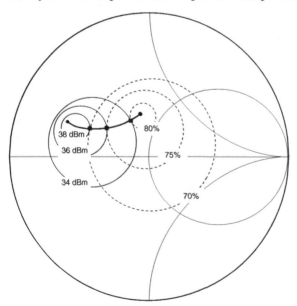

Figure 5.8 Load-pull contours for a 38 dBm GSM-900 DUT illustrating its optimum impedance trajectory and the optimum impedance state at 36 dBm and 75% PAE.

5.4.2 Simultaneous Average Load Power, PAE, and Signal Quality

The impedance state for optimum linearity can be identified with load-pull by looking at signal quality contours, including ACPR, ACLR, and EVM (error vector magnitude). The present approach expands on the previous method by adding signal quality contours, in addition to contours of average load power and PAE.

The general approach is to first identify the parameter, or parameters, that must be satisfied and rank them in order of importance. In general, since signal quality must always be satisfied, this would usually appear at the top of the ranking, followed by average load power,[6] then PAE.[7]

Consider the load-pull contours of Figure 5.9, where in addition to the usual average load power and PAE contours, contours for ACPR1 for WCDMA have been added. In general, there would be a pair of contours for ACPR1, but presently it assumed a wide-band low-memory test-fixture is implemented, and there is little to no symmetry. Assuming an ACPR1 specification of −50 dBc, it is seen that the −50 dBc ACPR1 contour is approximately orthogonal to the optimum impedance trajectory at approximately 23 dBm average load power and 46% PAE.

Now if it is assumed a minimum average load power of 24 dBm is required, then the process suggests that an associated ACPR1 of −51 dBc would obtain at the optimum impedance state, with an associated PAE of 44%. The process yields an approximate impedance state of $\Gamma_{load} \approx 0.4\angle 160°$.

Figure 5.9 Load-pull contours for a 26 dBm WCDMA DUT illustrating its optimum impedance trajectory and the optimum impedance state at 24 dBm and 44% PAE, with approximately −51 dBc ACPR1.

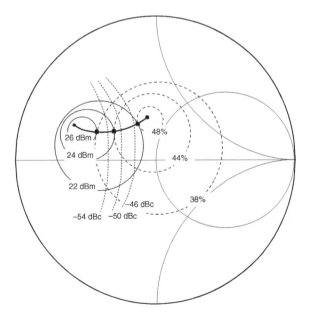

---

6 Since average load power is proportional to die area, average load power may, in fact, be a constraint, in which case the power density of the die becomes a factor in the load-pull process.
7 PAE is often the sacrificial specification.

5.4.3 Optimum Impedance Extraction Under Fixed-Parametric Load-Pull

Implicit in constant available source power is that transducer gain can, and will, change as the source and load impedance states are varied. In some applications, it is desired to hold transducer gain, or other dependent variables constant, which can only be achieved by fixed-parametric load-pull with iteration of available source power to hold the particular dependent variable constant, in effect making it an independent variable.

For example, consider the previous case of simultaneous optimization of average load power and PAE. Since available source power is an independent variable, and held constant, each load impedance state during load pull represents a different position on the instantaneous gain trajectory, and hence transducer gain and gain compression are each dependent variables. This is illustrated by the upper gain trajectory of Figure 5.10.

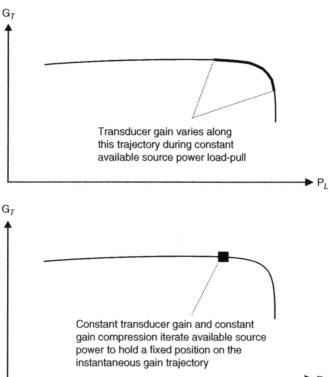

Figure 5.10 Instantaneous transducer gain trajectories illustrating transducer gain and gain compression displacement under constant available source power versus iteration of available source power to fix transducer gain and gain compression.

Consider now if, at each load impedance state, available source power is iterated to hold transducer gain and gain compression constant, as illustrated by the lower gain trajectory of Figure 5.10. Now the position on the instantaneous gain trajectory is fixed and other parameters can be explored and optimized, such as load power, for constant transducer gain and gain compression.

Constant parametric load-pull, such as constant transducer gain, has two advantages. First is that the effects of varying transducer gain and gain compression are removed. Second, and more important in optimum impedance extraction, is that transducer gain is effectively removed from the data, reducing its overall dimension by one. Thus, the entire chart Smith chart, subject to device under test (DUT) capability, represents contours for fixed transducer gain, thereby dramatically simplifying optimization, since two variables, instead of three, are plotted on the Smith chart.

5.4.4 PAE and Signal Quality Extraction Under Constant Average Load Power

To illustrate the simplest example of fixed-parametric load-pull, consider Figure 5.11, which illustrates the WCDMA example of Figure 5.9 with average load power fixed to 24 dBm. Note first the reduction in the number of contours, showing only ACPR1 and PAE contours. At any point along any contour, average load power is fixed to 24 dBm. In general, a DUT may not be able to exhibit the

Figure 5.11 Load-pull contours for a 26 dBm WCDMA DUT, under constant average load power load-pull, illustrating its optimum impedance trajectory and the optimum impedance state at −50 dBc ACPR1 and 48% PAE. An increase of 1 dB in ACPR1 has yielded a substantial 4% point improvement in PAE, from 44% to 48%, while reducing the parameter space from three to two.

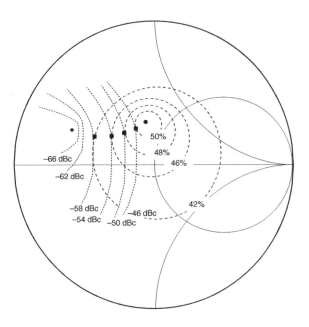

requested average load power, in which case the associated impedance domain must be adjusted. Impedance states included in the data not supporting the requested average load power will induce contour distortions unrelated to actual DUT behavior.

Second, observe that the -50 dBc contour now intersects the 48% PAE contour. It is evident that in the previous load-pull, the DUT was operating well below the maximum compression point, indicating that additional available source power could have been applied. Performing fixed available source power for various powers will achieve the present result, though being more complicated.

Using the optimum impedance state process, several optimum points are plotted in Figure 5.11, illustrating mutually tangent ACPR1 and PAE contours. The impedance state at -50 dBc and 48% is the desired solution, which delivers 24 dBm, yielding an improvement of 4%. The process yields an approximate impedance state of $\Gamma_{load} \approx 0.35\angle140°$, which makes sense, moving slightly closer to the maximum PAE impedance state.

5.4.5 Optimum Impedance Extraction with Bandwidth as a Constraint

The treatment up to now has implicitly assumed fixed single-frequency load-pull, with frequency being an implicit, but independent parameter. Because the vast majority of matching network applications span fractional bandwidth of the fundamental operating frequency, matching network synthesis and performance can be considered to extend to the present single-frequency graphical treatment to a narrow-band graphical method using constant-Q circles.

To traverse the bridge from single-frequency load-pull to narrow-band load-pull, consider Figure 5.12, which illustrates the familiar optimal trajectory of load power and PAE, for three different frequencies: low-band, mid-band, and high-band. The collection of each of the three impedance states[8] represents the corresponding optimal impedance trajectory versus frequency. Replication of this trajectory by the matching network yields, approximately, the designed optimal trade-off response over frequency, though some fine tuning may be done with an optimization run with Keysight Advanced Design System (ADS) to achieve the desired level of flatness.

To determine the effect of bandwidth as a constraint, consider the addition to Figure 5.12 a constant-Q circle of $Q = 1.5$, as illustrated by Figure 5.13. Note that all impedance states on the interior of the constant-Q circle do not constrain frequency response, for an associated matching network order. Each of the optimal

---

8 There is no reason to limit the frequency-domain trajectory to three impedance states, though three states are often chosen as a compromise between simplicity and accuracy.

Figure 5.12 Expanded-scale graphical representation of high-band, mid-band, and low-band optimal impedance trajectories, with each square denoting the desired optimal impedance at its corresponding frequency. Over frequency, the consolidated collection of impedance states becomes the optimal impedance trajectory over frequency, to be replicated by the matching network. In practice, the collection of optimal impedance states resembles Figure 4.17.

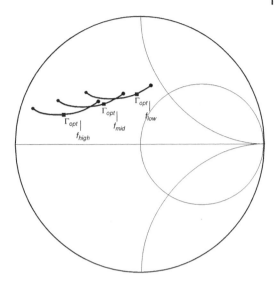

Figure 5.13 Extension of Figure 5.12 to illustrate the effect of narrow-band frequency response using a constant-Q circle of $Q = 1.5$. Note that the impedance states on the exterior of constant-Q circle require proportionally higher matching network order or, identically, a change in transistor power capability or technology. The optimal impedance states fall within the capture range of a matching network constrained by the $Q = 1.5$ circle, as illustrated.

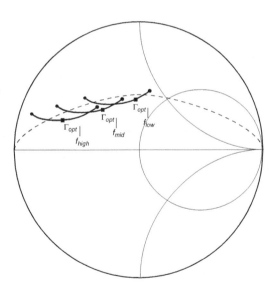

impedance states of Figure 5.13 lie on interior, and thus do not constrain frequency response. Because the scale is expanded to illustrate the concept of bandwidth as a constraint, the actual impedance states over frequency, in practice, would be substantially closer to each other. Figure 5.14 illustrates a similar bandwidth analysis for a 400 W PEP laterally diffused metal oxide semiconductor (LDMOS) DUT.

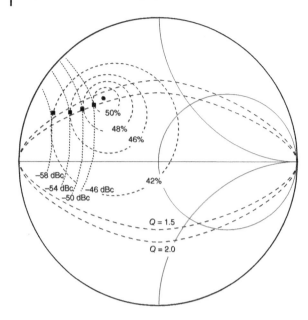

Figure 5.14 Contours of constant average load power of 43 dBm for a 400 W PEP LDMOS transistor illustrating ACLR1 and PAE contours superimposed over $Q = 1.5$ and $Q = 2.0$ circles for bandwidth analysis.

5.4.6 Extension to Source-Pull

Extension of the optimum impedance trajectory concept to source-pull is simple, with the load-pull contours simply replaced by source-pull contours and following the data acquisition methods of Chapter 4.

5.4.7 Extension to Harmonic and Base-Band Load-Pull

Extension to harmonic load-pull and base-band load-pull is slightly more involved, particularly since both methods often operate on fundamental optima already established and typically exhibit optima on the edge of the Smith chart. Substantially more iteration is therefore required, though the principles are the same.

5.5 Closing Remarks

The optimum impedance trajectory is the essential link between load-pull and design, embodying all the necessary external conditions that bestow optimal performance. Through systematic data acquisition under specific stimulus, automated load-pull yields sets of contours parametrized to impedance, frequency, bias, and power, rigorously describing complex nonlinear transistor behavior. Because stimulus and transistor embedding are realistic to arbitrary precision, each contour set represents a faithful embodiment of transistor behavior as a well-defined function of source and load impedance. Multiple contour sets plotted on the Smith chart provide a highly visual medium to identify explicit contour intersections, or regions of overlap, where underlying parameters are simultaneously optimum, possibly subject to additional constraints, such as bandwidth.

Explicit contour intersections, or regions of contour overlap, lead naturally to the concept of optimum impedance, which is defined as the impedance state that meets, or exceeds, all parametric objectives simultaneously. This is the impedance state that, when physically replicated by the matching network embedding the transistor, yields the optimal behavior identified by load-pull, net of matching network loss. The optimum impedance state is thus the key link between load-pull and optimal RF PA design.

6

Matching Network Design with Load-Pull Data

The foundations of matching network design evolved from classical two-port filter synthesis methods for telephone networks that were required to exhibit simultaneously a prescribed transmission response and conjugate impedance terminations at both ports for maximum power transfer [1–5]. Early theory proposed transmission responses that were either maximally flat in magnitude or phase, eventually being associated with Butterworth and Bessel–Thompson synthesis methods, respectively. Advanced development led to sharper roll-off response, as produced by Chebyshev and Cauer—Elliptical responses. Methods based exclusively on distributed-network synthesis were also proposed, including the Klopfenstein and Hecken tapers, each proposing minimal-length tapers for given impedance transformation and bandwidth.

For wireless power amplifier (PA) design, the matching network is required to present an impedance termination, at all relevant mixing products, that delivers the desired performance identified by load-pull. At the physical level, several things are happening concurrently. Because the supply voltage a transistor operates at may not necessarily be the optimum voltage of the load, the matching network can be viewed as impedance inverter that simultaneously allows maximum voltage swing across the load and maximum voltage swing across the transistor, subject to supply voltage or breakdown voltage limitations. The physical response of the matching network can also be interpreted as a means to deliberately manipulate the relative amplitude and phase of voltage and current at the transistor to achieve some desired goal, such as maximum symmetrical voltage swing for optimum power or maximum small-signal gain for maximum power-added efficiency (PAE).

The matching network is a central component of the load-pull method of PA design. It represents the link between the terminating impedances identified by the methods of Chapter 5 and the target impedances that must be synthesized by the matching network. The matching network impedance trajectory is not unique in that there are many responses available to generate a specified impedance, with

The Load-Pull Method of RF and Microwave Power Amplifier Design, First Edition. John F. Sevic.
© 2020 John Wiley & Sons, Inc. Published 2020 by John Wiley & Sons, Inc.

each response offering relative merit with respect to performance, physical size, and implementation cost.

6.1 Specification of Matching Network Performance

Impedance transformation ratio, bandwidth, insertion loss, harmonic rejection, and physical size are the primary factors establishing impedance transformation network performance. Secondary factors include out-of-band response, especially for stability control and video bandwidth (VBW) control, and bill of materials (BOM) and assembly cost. The number of sections is fixed by the transformation ratio, bandwidth, and insertion loss and is the primary contributor to overall cost as size is directly proportional to the number of sections.

Matching network design consists of the following steps, each of which are treated presently.

- Appraisal of the matching network specification for each stage
- Establish network topology and number of sections
- Select physical synthesis method from lumped, distributed, or hybrid
- Graphical or CAD optimization
- Physical implementation
- Fabrication and evaluation

6.2 The Butterworth Impedance Matching Network

For narrow-band applications encountered in the commercial wireless sector, the multi-section Butterworth impedance matching network shown in Figure 6.1 represents an optimum compromise in bandwidth, insertion loss, ripple, harmonic rejection, and ease-of-use.[1] While other responses offer advantages such as increased harmonic rejection, for example with the Chebyshev response, these come with disadvantages, like degraded group delay and physical implementation difficulty. For applications that require matching bandwidth approaching a decade, such as the harmonic load-pull test-fixture, optimum distributed-network tapered-lines are used, the most popular being the Hecken matching network.

The Butterworth matching network of Figure 6.1 will match R_2 to R_1, their ratio defined as the impedance transformation ratio, T. The bandwidth over which

---

1 The Butterworth matching network belongs to the class of maximally flat networks, which is equivalent to geometric-mean matching. It is this property of the Butterworth matching network that makes it amenable to graphical construction with the Smith chart while simultaneously offering adequate bandwidth with reasonable group delay, critical for optimum VBW.

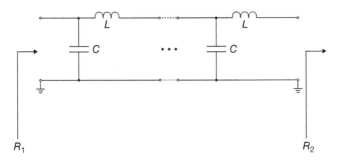

Figure 6.1 An *N* section matching network with overall impedance transformation ratio *T*. The impedance transformation ratio of each section is T_p. For the Butterworth matching network, T_p of each section is identical, so that $T = T_p^N$.

impedance matching occurs is directly proportional to the number of sections, *N*, which can vary from one for a GSM hand-set RF PA interstage match, to five for an laterally diffused metal oxide semiconductor (LDMOS) input pre-match of an infrastructure RF PA. Figure 6.1 also illustrates the standard convention of describing the impedance transformation the matching network is designed to synthesize, launching from impedance R_1 to impedance R_2.

6.2.1 The Butterworth *L*-Section Prototype

Figure 6.2 shows one of four canonical *L*-section prototypes to which Butterworth filter synthesis can be applied to design each section of the matching network of Figure 6.1. Aside from the advantages just mentioned, its central merit is its relative ease of synthesis and physical implementation. Since the Butterworth response is equivalent to geometric-mean matching, the entire synthesis and design process can be carried out exclusively by simple closed-form analytical expressions or graphical methods based on the Smith chart.

Figure 6.2 Low-pass Butterworth *L*-section matching prototype.

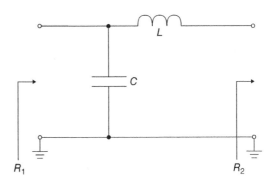

The remaining three *L*-section prototypes are illustrated in Figure 6.3 and represent the class of fundamental matching network building blocks capable of synthesizing an arbitrary impedance anywhere on the interior of the Smith chart. Each *L*-section prototype is distinguished by either a low-pass or high-pass frequency response and the relative direction of its impedance transformation. Associated with each prototype is a unique impedance trajectory, graphically illustrating its distinct properties in achieving the performance requirements specified by the matching network appraisal discussed in Section 6.1.

The low-pass impedance trajectory provided by the *L*-section prototype of the upper-left corner of Figure 6.3, well serves matching network requirements for the vast majority of RF PA applications. As an output matching network, two or three sections will often be cascaded to generate a low impedance load-line at the required bandwidth. The input matching network follows a similar topology, whereas the interstage matching network is usually high-pass or mixed, as illustrated by the *L*-section prototype in the lower-left corner. The high-pass *L*-section prototype also presents a DC block, which is required by interstage matching to isolate each bias network.

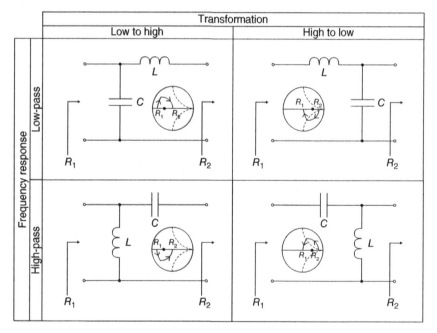

Figure 6.3 The four canonical Butterworth *L*-section matching sections based on frequency response and transformation. The solid dot on the real axis is the geometric mean between R_1 and R_2 and represents the characteristic impedance of the quarter-wave line matching these two impedances.

The first step of Butterworth network synthesis is determination of N, the number of sections necessary to achieve its target specifications. Since the impedance transformation ratio, T_p of each L-section prototype of the Butterworth matching network is identical, the overall impedance transformation ratio is

$$T = T_p^N = \frac{R_2}{R_1} \tag{6.1}$$

where R_2 and R_1 are the two impedances being matched, as shown in Figure 6.1. The intermediate impedance transformation ratio is defined as

$$T_p = \frac{R_2}{R_1} \tag{6.2}$$

where R_2 and R_1 are defined in Figure 6.2. Unique resolution of N is constrained by bandwidth, which requires identification of an additional (generally non-linear) equation in N and bandwidth, enabling construction of a 2×2 system of equations yielding a unique solution for N at a specified bandwidth and impedance transformation ratio, T. Solution of this nonlinear system can be carried out by analytical or graphical methods, next described.

6.2.2 Analytical Solution of the Butterworth Matching Network

An analytical closed-form solution for N is identified by augmenting Eq. (6.1) with an additional equation in T and instantaneous bandwidth. The solution process can be simplified by assuming an empirically optimum proxy for instantaneous bandwidth, matching network Q, so that for T specified by the matching network appraisal, the required number of matching section's, N, results. From N, L, and C of each L-section prototype can be calculated, concluding the synthesis process. From these L-section prototypes, lumped-parameter, distributed-parameter, and hybrid-parameter matching networks are then physically implemented.

Though subject to the goals of a particular matching network design, it has been observed by the author from a substantial history of empirical evidence, that setting $Q_o \leq 2.0$ of the Butterworth L-section renders it well-suited for an RF PA instantaneous BW 4–5% of the operating frequency. Ultra-broadband multi-carrier power amplifier (MCPA) applications employing third-order digital pre-distortion (DPD) may require $Q_o \leq 1.5$, providing suitable performance to an instantaneous bandwidth of 10% of the operating frequency. The number of matching sections required for a given transformation ratio is approximated by solving

$$\sqrt[N+1]{T-1} \leq Q_o \tag{6.3}$$

for N rounded up to the next largest integer and Q_o being set to either 2.0 or 1.5, depending on the application, N representing the number of individual cascaded

Butterworth L-section prototypes, and T the impedance transformation ratio. Since N must be integer-valued, solution of Eq. (6.3) is easily solved through iteration, from which L and C follow. From the standard definition of network Q, L and C of the L-section Butterworth prototype are

$$L = \frac{Q_o R_L}{\omega_0} \tag{6.4}$$

$$C = \frac{1}{\omega_0^2 L} \tag{6.5}$$

where R_L and R_H are defined in Figure 6.2 and ω_0 is the resonant radian frequency of the matching network.

Through a series of algebraic manipulations, expressions for L and C can be derived for an arbitrary number of sections greater than one. This exercise will be left to the reader. Alternatively, CAD filer synthesis tools can be used, such as Keysight Advanced Design System (ADS).

The analytical treatment has transformed between real-valued impedances, though this is not a requirement or limitation. Adding an imaginary component to the matching network is effected by simply adding a series or shunt reactive element to the Butterworth L-section, though its best done graphically, as the effect on bandwidth is easier to assess, as Section 6.2.3 demonstrates.

6.2.3 Graphical Solution of the Butterworth Matching Network

Constant-Q circles,[2] establish an elegant duality between graphical solution of the Butterworth L-section matching network and its analytical equivalent using the closed-form solution given by Eq. (6.3). Instead of analytical calculation of N, the graphical method indirectly identifies the number of sections geometrically by bounding the absolute value of the instantaneous deviation of the impedance trajectory to lie between the specified Q_o circle and the real-axis, as illustrated in Figure 6.4 for a two section matching network. Since the Butterworth L-section is equivalent to geometric mean matching, the geometric mean of the impedance loci intersections with the real-axis represents the characteristic impedance of the quarter-wave line matching these two impedances, providing a direct route to distributed-parameter synthesis. These two associated characteristic impedances are shown in Figure 6.4.

Graphical solution of L and C of a matching network required to match R_1 to R_2 is illustrated by Figure 6.4, with a design target specified by a $Q_o = 2.0$ circle. Geometrically, the solution is interpreted as inserting series inductive reactance from R_1 until it intersects with the $Q_o = 2.0$ circle, followed by insertion of shunt capacitive susceptance that intersects with the real-axis at R_a, followed by an additional

---

2 Geometrically, they are not circles, but often colloquially referred to as constant-Q circles.

Figure 6.4 Impedance displacement loci for construction of a two-section Butterworth distributed-parameter matching network with $Q_o = 2.0$. Note that impedances Z_1 and Z_2 are the geometric means of the impedance pairs R_1 and R_a and R_a and R_2, respectively, and thus represent characteristic impedance of the quarter-wave lines matching these impedance pairs.

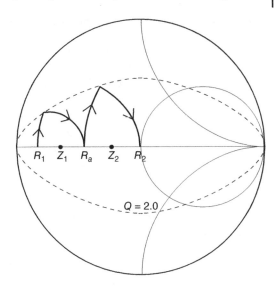

L-section composed of a series inductor and shunt capacitor, again bounded by the $Q_o = 2.0$ circle. For the relative values of R_1 and R_2 shown, this matching network generates the required matching and bandwidth with two sections, as the impedance loci illustrate. Where R_1 to be substantially lower, for example, the first inductance loci would intersect with the $Q_o = 2.0$ circle sooner, thus requiring insertion of a shunt capacitance, thereby adding an additional section to maintain the equivalent bandwidth between the two matching networks with substantially different impedance transformation ratios. This is an entirely general trend. Following calculation of L and C from Smith chart operations, Keysight ADS may be employed for additional refinement and optimization, particularly over bandwidth, or tables may be used [6, 7].

6.3 Physical Implementation of the Butterworth Matching Network

Matching network synthesis consists of selecting a frequency response, transformation direction identification, and calculation of the number of sections for the required bandwidth and impedance transformation ratio, yielding the lumped-parameter constants L and C of each L-section prototype. From the L-section prototype follows the procedure of physical synthesis at the circuit level, based on the lumped-parameter, distributed-parameter, or hybrid-parameter implementations.

6.3.1   The Lumped-Parameter Butterworth Matching Network

Physical implementation of the lumped-parameter Butterworth matching network follows directly from synthesis since both are expressed in terms of L and C, so no further rendering of network parameters is required. Each lumped element identified by the synthesis procedure is mapped directly into its corresponding element of physical implementation of Figure 6.1.

An important consideration for lumped-element matching network implementation at wireless frequencies is parasitic phenomena. Since the wavelength at wireless frequencies is on the order of the physical dimensions of most passive components used for the physical implementation of the matching network, deliberate scrutiny must be attached to their parasitic resonances and losses, their physical orientation, and the physical environment in which they are embedded, such as the substrate and physical placement that can induce magnetic field coupling to adjacent components. Resonance phenomena can also be deliberately exploited, such as series resonance of the capacitor as a DC block and parallel resonance of the inductor as an RF choke or bias feed. Several examples are provided later illustrating the importance of this effect.

Lumped-parameter matching networks are appropriate for HF and VHF applications. At the common wireless bands spanning 700–2100 MHz, lumped-parameter matching networks have the advantage of physical compactness, especially inductance, though typically have higher losses than distributed-parameter networks of equivalent operating frequency and impedance transformation. At these frequencies, hybrid implementation is usually optimum.

Example 6.1   Design a lumped-parameter Butterworth matching network to match 50 to 5 Ω for a GSM-450 final stage spanning 450–460 MHz. Use the graphical method and adopt a low-pass topology for harmonic suppression.

Solution
First, select the low-pass L-section prototype of the upper-left corner of Figure 6.3 as it represents the appropriate direction of impedance transformation and is the low-pass topology. Then, using the Smith chart of Figure 6.4, $R_1 = 5\ \Omega$ and $R_2 = 50\ \Omega$, corresponding to the impedance definitions of Figure 6.1. From R_1, addition of series inductance of $L_1 = 3.3$ nH and shunt capacitance of $C_1 = 27$ pF complete the first section, with an intermediate transformation to $R_a = 22.7\ \Omega$. Launching off the intermediate impedance, R_a, a second series inductance of $L_2 = 8.5$ nH and shunt capacitance of $C_2 = 7.8$ pF are added between the real-axis and the $Q_o = 2.0$ boundary, thus completing the two L-section prototypes. The Smith chart operations were carried out at 455 MHz as the midpoint of the lower and upper frequencies.

Figure 6.5 Physical implementation of the two-section low-pass Butterworth matching network matching 50 to 5 Ω for Example 6.1.

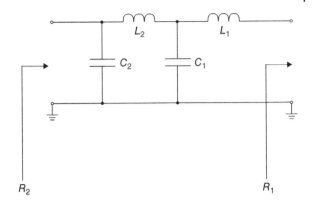

There is a one-to-one mapping between each parameter of the lumped-parameter prototype and its associated physical implementation. Thus, the completed two section Butterworth matching network is illustrated in Figure 6.5. ADS simulation allows further frequency response optimization while adoption of accurate passive models would allow parasitic effects to be analyzed. For example, series resonance of a chip capacitor increases its apparent capacitance, causing the capacitor to overshoot its intersection with the real axis. Similarly, parallel resonance of a chip inductor also increases its apparent inductance, causing an overshoot of the constant-Q circle boundary. Simulation, with accurate models simplifies analysis of parasitic effects to enable to correct passive component values to be used that correctly approximate the Butterworth synthesis in the physical implementation.[3]

Example 6.2 Quantify the effect of parasitic phenomena on a single-section Butterworth lumped-parameter matching network, using thin-film chip components, by assessing the apparent reactance of each element and comparing the result to ideal chip components. Assume $Q_0 = 2.0$.

Solution
To first-order, a chip inductor exhibits parallel parasitic resonance by interwinding capacitive coupling. Similarly, a chip capacitor exhibits series parasitic resonance by electrode inductance exhibited by each plate. Chip component vendors often specify parasitic resonance effects by specifying the frequency at which resonance occurs. Less frequent is specification of an equivalent lumped-parameter value of parasitic capacitance or inductance, though this approach requires geometric effects to be specified, for example, for different component sizes.

---

3 Modelithics provides an excellent library of well-validated passive and active models.

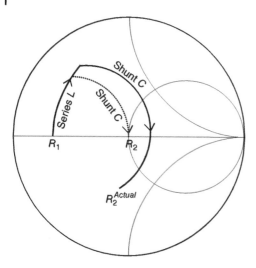

Figure 6.6 Impedance loci of a single-section low-pass Butterworth matching network for ideal chip components (dashed line) and chip components with first-order parasitic resonance effects (solid-line). Note that for both inductance and capacitance, the effect of parasitic resonance is to increase the apparent reactance from nominal, causing overshoot of each loci from the intended target.

Consider a 0603 chip capacitor with a nominal capacitance of 3.3 pF and series resonance at 3.4 GHz. Similarly, consider a 0603 thin-film chip inductor with nominal inductance of 1.7 nH and parallel resonance at 4.1 GHz. Ignoring parasitic resonance, this low-pass single-section Butterworth matching network will match 10–50 Ω at 1.9 GHz, as illustrated by the ideal impedance loci in Figure 6.6.

Using the parasitic resonant frequencies provided, the apparent inductance is 2.2 nH and apparent capacitance is 4.8 pF, yielding the impedance also illustrated in Figure 6.6, yielding an impedance of $17 - j33\,\Omega$. Although this example was chosen deliberately to exaggerate parasitic resonant effects, it is nevertheless evident parasitics have an effect and must be accounted for.

6.3.2 The Distributed-Parameter Butterworth Matching Network

Physical implementation of the multi-section distributed-parameter Butterworth matching network is a subset of lumped-parameter synthesis, since the characteristic impedance of the former is derived graphically from the lumped-parameters of the latter. To physically implement single-stage distributed-parameter matching network simply requires identification of the characteristic impedance of the quarter-wave transmission line matching two real-valued impedances R_2 and R_1, expressed as

$$Z_0 = \sqrt{R_1 R_2} \tag{6.6}$$

where Eq. (6.3) has been satisfied for $N = 1$. Extension of Eq. (6.6) to match reactive elements is accomplished by augmenting the electrical length of the transmission line to provide the conjugate match, subject to the loci being bounded by the

$Q_o = 2.0$ circle to maintain the condition for $N = 1$. If the loci extends beyond this region, additional sections are required.

If graphical analysis shows more than one section is required, or if Eq. (6.3) shows two or more sections are required to generate the impedance transformation ratio, T, at the specified bandwidth, then distributed-parameter synthesis is treated first as a lumped-parameter synthesis problem to solve for L and C of each section. From these lumped-parameters, the characteristic impedance of each section is graphically synthesized as illustrated in Figure 6.4 for a two-section Butterworth matching network. The characteristic impedance of each section of the distributed-parameter network is established by the geometric mean of each pair of adjacent real-axis impedance intersections, illustrated by Z_1 and Z_2 in Figure 6.4 where $Z_1 = \sqrt{R_1 R_a}$ and $Z_2 = \sqrt{R_a R_2}$. This process can be extended to additional sections by simply calculating the characteristic impedance from each pair of intermediate real-axis impedances. Extension of the multi-section matching network to matching reactive elements is treated the same as it is with the single-section matching network.

Example 6.3   Physically implement a distributed-parameter matching network to match 5.5–50 Ω from 1.710 to 1.785 GHz, to be used as a DCS up-link fundamental load-pull test-fixture. Use the graphical method.

Solution
The impedance transformation ratio, T, is 9. Since the bandwidth is approximately 5% of the operating frequency, $Q_o = 2.0$ is adopted as the synthesis boundary condition, with Eq. (6.3) yielding $N = 2$ sections. Similarly, using the Smith chart of Figure 6.4 with $R_1 = 5.5\ \Omega$ and $R_2 = 50\ \Omega$, it is found graphically that $L_1 = 0.7$ nH, $C_1 = 8.0$ pF, $L_2 = 2.2$ nH, and $C_2 = 2.7$ pF, from which the intermediate impedance is found to be $R_a = 16.7\ \Omega$. From the intermediate impedance, the characteristic impedances of each section are $Z_1 = 9.5\ \Omega$ and $Z_2 = 28.9\ \Omega$.

Figure 6.7 shows the physical implementation of the two-section distributed matching network specified in electrical length. Transmission line synthesis tools

Figure 6.7 Physical implementation of the distributed-parameter matching network for Example 4.3 based on a two-section lumped-parameter L-section synthesis. $Z_1 = 9.5\ \Omega$ and $Z_2 = 28.9\ \Omega$.

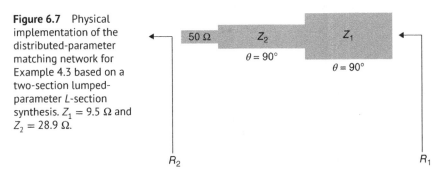

can be used to design the physical length for a given wave-guiding structure and material constants, though micro-strip is the most common.

6.3.3 The Hybrid-Parameter Butterworth Matching Network

Contemporary wireless standards span 400 MHz to 3 GHz at the fundamental, reaching 5.7 GHz if WiFi is included. As the three previous examples illustrated, spanning this frequency range, lumped-parameter and distributed-parameter implementations each exhibit mutually exclusive benefits, such as compact physical size and minimal parasitic effects, that if combined would yield an optimum physical implementation. The hybrid-parameter implementation is such an optimum implementation, combining the flexibility and parasitic performance of distributed inductance with ease-of-tuning and compact physical size of lumped capacitance.

Synthesis of the hybrid-parameter matching network exploits the unique ability of the Smith chart to simultaneously resolve lumped-parameter and distributed-parameter operations. In the most common implementation, a series transmission line section is loaded by shunt lumped capacitance to replicate one or more low-pass *L*-section prototype sections of Figure 6.2. Figure 6.8 illustrates a canonical two-section hybrid-parameter matching network, demonstrating its distinct advantage of post-fabrication ease-of-tuning by allowing real-time movement of shunt capacitance and rapid alteration of the characteristic impedance and electrical length of the transmission line using the dot-array. These two tuning mechanisms enable both fine and coarse adjustments to be made to precisely replicate the target impedances obtained by load-pull. Combined with a laser and

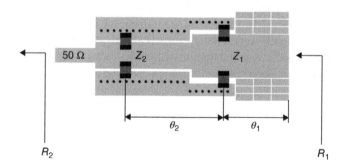

Figure 6.8 Physical implementation of the hybrid-parameter matching network illustrating the dot-array to solder copper tape for changing characteristic impedance and electrical length and top-metal via arrays for soldering chip capacitors an arbitrary electrical length equivalent to series inductance. The topology illustrated is electrically equivalent to the two-section lumped-parameter and distributed matching networks illustrated in Figures 6.5 and 6.7, respectively.

proper metal-mask layout, the underlying principle of post-fabrication tuning can be applied to MMIC design as well.

Physical implementation of the hybrid-parameter matching network combines elements of both lumped-parameter and distributed-parameter methods, and is best carried out using a graphical solution with the Smith chart. To begin, consider the two-section hybrid-parameter Butterworth matching network of Figure 6.8 whose transformation loci is illustrated in Figure 6.9, with starting impedance, R_1. This impedance is located on the Smith chart, with its angular displacement along its constant-resistance loci to the intersection with the Q_o circle defined as θ_1 of Figure 6.9. From this intersection, the process then proceeds in a fashion identical to lumped-parameter synthesis by adding a shunt capacitance whose loci intersects the real-axis at R_a. Launching from R_a along its constant-resistance loci to the intersection with the Q_o circle forms the angle θ_2, illustrated in Figure 6.9. Addition of shunt capacitance along its constant-conductance loci to R_2 completes the physical implementation. Note that parallel grouping of the chip capacitors, a common practice, decreases their heating loss contribution by a factor of 2 while increasing their combined effective resonant frequency by 41%. Each chip capacitor would then be approximately one half their value specified by synthesis, depending on how close the operating frequency is to their first resonance.

The Smith chart operations to implement the matching network of Figure 6.8 are illustrated in Figure 6.9, particularly the micro-strip angular displacement definitions. A small angular displacement is approximately equivalent to a series inductance displacement along a constant-resistance loci, with the deviation from the

Figure 6.9 Impedance displacement loci for construction of a two-section low-pass Butterworth hybrid-parameter matching network physically implementing Figure 6.8 by distributed series inductance and lumped shunt capacitance.

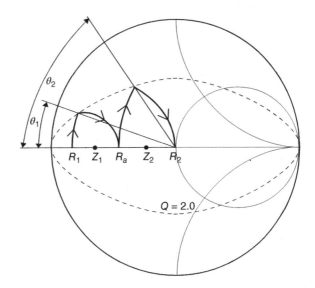

constant-resistance loci established by the characteristic impedance of the transmission line. The transmission line characteristic impedance of each section of the hybrid-parameter matching network is an exogenous variable that is chosen, most often, from physical criteria of the matching network, such as transistor lead width. In contrast, with lumped-parameter and distributed-parameter synthesis, the characteristic impedance of each transmission line section is an endogenous parameter calculated from lumped-parameter constants L and C.

Example 6.4   Design a PCS infrastructure PA hybrid-parameter matching network for internally pre-matched LDMOS to match 5–50 Ω from 1930 to 1990 MHz. Adopt a low-pass topology for harmonic suppression and assume the substrate is Rogers 4350 with a nominal dielectric constant of 3.48 and thickness of 50 mils. Specify the line length and width of each transmission-line section in electrical length as well as physical length, the electrical and physical location of each capacitor with respect to the low impedance side of the network, and capacitor values assuming infinite resonant frequency. Assume DPD is not used and assume the lead width of the transistor is 400 mils.

Solution
The solution to this example will illustrate the earlier point about the characteristic impedance of the transmission line sections being exogenous variables, meaning that there is no closed-form solution or graphical method to identify their values. Instead, the characteristic impedances are subject to engineering judgment centered around physical implementation requirements, such as transistor lead width and substrate thickness for a reasonable characteristic impedance.

The impedance transformation ratio, T, is 10. Since the bandwidth is 60 MHz and DPD is not used, $Q_o = 2.0$ is adopted as the synthesis boundary condition, with Eq. (6.3) showing that $N = 2$ sections are expected to generate this impedance transformation ratio at this bandwidth, though it is at the very edge of meeting the bandwidth criteria.

To prevent discontinuity effects at the matching network and transistor lead interface, choosing the micro-strip width of the first section to be 400 mils is a reasonable starting point. Based on this width, the substrate thickness, and the dielectric coefficient of Rogers 4350 substrate, its characteristic impedance is 20 Ω. Figure 6.10 illustrates the impedance loci associated with identifying θ_1 and C_1 using a Smith chart with a reference impedance of 10 Ω. These values are 20° and 8 pF, respectively, with each capacitor value of the pair 4 pF nominal.

Using $R_a = 13 \ \Omega$ as the launching impedance of the second stage leads to the impedance trajectories illustrated by Figure 6.11, where the transmission line impedance has been selected to be 50 Ω since sufficient transformation can be provided by this value coupled with the expected shunt capacitance. Where

Figure 6.10 Impedance transformation loci of the first section of the hybrid-parameter Butterworth matching network of Example 4.4. The reference impedance of the Smith chart is 10 Ω.

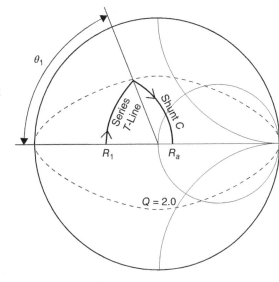

Figure 6.11 Impedance transformation loci of the second section of the hybrid-parameter Butterworth matching network of Example 4.4. The reference impedance of the Smith chart is 50 Ω.

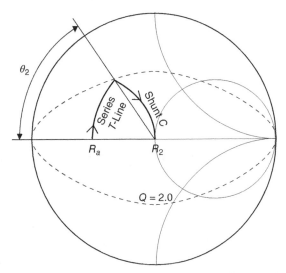

additional transformation required, or the chip capacitor values small, meaning sensitive to placement and manufacturing variation, a lower characteristic impedance could be used. From the Smith chart loci shown in Figure 6.11, θ_2 and C_2 are found to be 32° and 2.5 pF, respectively. Given the relatively small value of capacitance, it is prudent to use only one capacitor for shunt susceptance. Given that this is the second section of the matching network, current is low and shunt capacitance loss will be minimal.

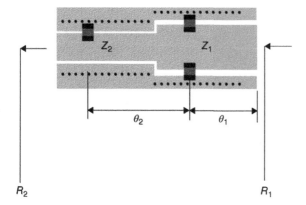

Figure 6.12 Physical implementation of the hybrid-parameter two-section low-pass Butterworth matching network of Example 4.4.

Figure 6.12 shows the physical implementation of the matching network design with physical displacement of the capacitors converted from electrical length at the center of the design band. The physical length of the first section is 200 mils and the physical displacement of the second shunt capacitor from the first is 332 mils. For a characteristic impedance of 50 Ω for the second section, a line width of 132 mils is required. Note that the primary effect of the hybrid-parameter implementation is lumped capacitance shortens the transmission lines while providing the advantage of real-time tuning. This implementation is by far the most popular of modern PA implementations.

6.4 Supplemental Matching Network Responses

The hybrid-parameter Butterworth matching network is suitable for most RF PA applications, spanning the GSM handset to MCPA architectures with DPD. It can deliver a reasonably flat impedance up to 10% operating bandwidth, where it then becomes impractical due to the number of sections required. In those applications where either additional impedance transformation or bandwidth is necessary, the Chebyshev and Hecken responses are preferred. Because the Hecken response is capable of a decade bandwidth, it is ideal for load-pull test-fixture applications where harmonic impedance transformation and fundamental frequency agility are necessary.

The Bessel–Thompson response is the dual of the Butterworth response for phase, thereby offering maximally flat group delay. It is often used in bias network design where memory effects must be minimized. Each of these responses is discussed presently.

6.4.1 The Chebyshev Response

The Chebyshev matching network would typically be used in applications where physical constraints prevent the use of the Butterworth implementation, for a given impedance transformation or bandwidth. For equivalent performance impedance transformation or bandwidth, it is slightly smaller physically, though it is subject to increased manufacturing sensitivity, particularly chip capacitor placement. The Chebyshev network also suffers from degraded group delay compared to Butterworth and Hecken responses.

Because the Chebyshev response is not geometric mean, it is not possible to implement a solution using the Smith chart. Tables or numerical methods are used, based on specifying the usual parameters from Butterworth synthesis, including impedance transformation ratio and bandwidth. From the tables, along with a new parameter called in-band *VSWR*, tabulated solutions provide the number of sections and the associated normalized characteristic impedance of each section for a specified impedance transformation ratio and bandwidth. From this prototype, lumped-parameter, distributed-parameter, and hybrid-parameter matching networks can be physically implemented. Alternatively, numerical methods can be used in place of the tables, the process and results being the same. The standard references for Chebyshev matching network synthesis are [6, 8], while Keysight ADS can be used for numerical synthesis.

Example 6.5 Determine the number of sections for a Chebyshev matching network for a PCS downlink system spanning 1930–1990 MHz with third-order DPD correction. Assume an impedance transformation ratio of 50. Calculate the number of sections for an equivalent Butterworth network and compare.

Solution
The fundamental frequency spans 1930–1990 MHz; with third-order DPD the operating BW is therefore 180 MHz. For margin and degraded group delay performance of the Chebyshev response, round up to 200 MHz operating bandwidth, giving a normalized bandwidth of 0.1.

The additional variable for Chebyshev synthesis is in-band *VSWR*, for which 1.05 is assumed. From this specification, table 6.02-4 of Matthaei yields $N = 3$. For an equivalent Butterworth implementation, Eq. (6.3) shows that nine sections would be required for similar performance, indicating a substantial improvement with Chebyshev synthesis.

6.4.2 The Hecken and Klopfenstein Responses

The Hecken matching network belongs to the class of distributed-networks distinguished by a continuous, generally nonlinear taper, yielding the shortest possible electrical length for specified impedance transformation and bandwidth [9–11]. In contrast, maximally compact distributed-networks such as Dolph–Chebyshev and Klopfenstein rely on symmetrical discontinuities that although yield the shortest possible electrical length for specified impedance transformation and bandwidth result in impaired group delay [12].[4] Given the fundamental problem that the Hecken taper solved – what is the minimum discontinuity-free electrical length necessary for a specified impedance transformation ratio and bandwidth – it is somewhat surprising how recent the solution was published compared to other responses.

The Hecken taper is most commonly used for load-pull test-fixture pre-matching networks where its modest impedance transformation ratio is traded off for its exceptional operating bandwidth, which can exceed a decade, and its optimum group delay performance. Expanded bandwidth is necessary for multi-harmonic load-pull and instantaneous frequency agility that is impossible with individual quarter-wave sections or difficult with multi-section transformers. Constant group delay subjects each individual component of a wideband modulated spectrum to uniform group velocity, eliminating spectral asymmetry that can impair adjacent channel power ratio (ACPR) and ACLR characterization. Nonuniform group velocity over the modulation bandwidth is one form of memory common in load-pull, of several, that must be rigorously controlled to reduce as much as possible any influence the load-pull system or test-fixture might have on signal quality.

As a continuous distributed-network, solution of the Hecken taper is based on resolving its instantaneous characteristic impedance trajectory versus physical displacement, in contrast to solution of discrete L-sections for Butterworth design. Because the solution involves transcendental functions, numerical methods are often adopted to resolve the instantaneous characteristic impedance trajectory. Design of the Hecken taper of Figure 6.13 begins in the usual manner by specifying

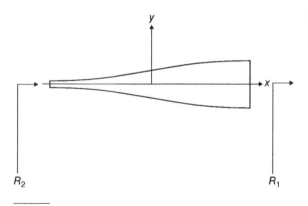

Figure 6.13 A single-section Hecken taper matching network.

---

4 The optimum Hecken taper is at most 14% longer than an equivalent Klopfenstein taper.

Figure 6.14 A generic Hecken taper illustrating instantaneous impedance response versus physical length.

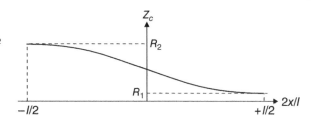

an impedance transformation ratio and an associated operating bandwidth. The Hecken taper solution is specified by its instantaneous characteristic impedance trajectory, $Z_c(\xi)$, shown in Figure 6.14, defined as

$$2 \ln Z_c(\xi) = \ln(Z_1 Z_2) + \ln\left(\frac{Z_2}{Z_1}\right) G(B, \xi) \tag{6.7}$$

where $G(B, \xi)$ is defined as

$$G(B, \xi) = \frac{B}{\sinh B} \int_0^{\xi} I_0(B\sqrt{1 - \zeta^2})d\zeta \tag{6.8}$$

with $I_0(z)$ the modified Bessel function of first kind and $\xi = \frac{2x}{l}$ with x instantaneous physical displacement along the Hecken section, as defined in Figure 6.14, and l physical length, and $|x| < \frac{l}{2}$.

Although the instantaneous impedance, Z_c, is a continuous function of electrical length, approximate design using planar electromagnetic (EM) simulation tools, such as Sonnet, begins by discretizing Z_c. A reasonable starting point for discretization on micro-strip is limiting each rectangular element no larger than $\lambda/10$ followed by filling in each element with quasi-triangular elements to approximate a continuous taper. From Sonnet, the two-port s-parameters can be simulated for an arbitrary substrate to verify the impedance transformation ratio and group delay, as well as frequency response and insertion loss. The accuracy delivered by this method, coupled with the robustness of Sonnet, will yield first-pass success. The following example illustrates numerical solution of Eq. (6.7), which is implemented on an alumina thin-film substrate.

Example 6.6 Evaluate Eq. (6.7) to provide a 5 : 1 impedance transformation for a load-pull test-fixture spanning 800 MHz to 8.0 GHz. Express the instantaneous impedance Z_c as a function of electrical length by plotting it in rectangular coordinates. Plot R_1 in the gamma-domain. Assume a pass-band ripple of −40 dB.

Solution
Evaluation of Eqs. (6.7) and (6.8) requires nonlinear solution of the former and numerical integration of the latter. Mathworks MATLAB provides a convenient

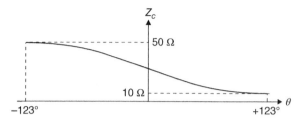

Figure 6.15 Hecken instantaneous characteristic impedance trajectory for Example 6.6. The electrical length is at the nominal design frequency.

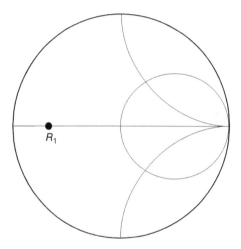

Figure 6.16 Hecken matching network input impedance spanning 800 MHz to 8 GHz for Example 7.6. The Smith chart reference impedance is 50 Ω.

platform to carry out these tasks by providing robust nonlinear solvers and numerical integration algorithms.[5]

Implementing Eqs. (6.7) and (6.8) in MATLAB yields the instantaneous impedance trajectory, Z_c, shown in Figure 6.15. The transformation is from 10 to 50 Ω spanning 800 MHz to 8 GHz, with R_1 plotted on the Smith chart shown in Figure 6.16. Note the exceptionally flat response over this bandwidth, being essentially a dot, meaning fundamental and harmonics each receive uniform impedance transformation while offering frequency agility and constant group delay.

---

5 Though there are many different numerical analysis packages, The Mathworks MATLAB platform is highly recommended for its comprehensive library of algorithms, graphics capability, and the robustness of its routines, especially the nonlinear solvers.

6.4.3 The Bessel–Thompson Response

The Bessel–Thompson response is the phase response dual of the Butterworth maximally flat magnitude response, providing maximally flat phase response, equivalent to constant group-delay. This property is attractive for use in wideband bias network synthesis where exceptionally high VBW is necessary to minimize linear distortion and its manifestation as IM and ACPR asymmetry. The Bessel–Thompson response is commonly used in MCPA applications and, with the Hecken taper, for ultra-wideband load-pull test-fixtures for evaluating linearity spanning 200–300 MHz VBW under DPD correction.

Tables or numerical methods are used for Bessel–Thompson synthesis, based on specifying the usual parameters from Butterworth synthesis, including impedance transformation ratio and bandwidth. From the tables, along with an additional parameter describing maximum allowable group delay, the tables provide the number of sections and the associated normalized characteristic impedance of each section. From this prototype, lumped, distributed, and hybrid matching networks can be physically implemented. Alternatively, numerical methods can be used in place of the tables, the process and results being the same. The standard reference for Bessel–Thompson matching network synthesis is [6, 13], while Keysight ADS can be used for numerical synthesis.

6.5 Matching Network Loss

6.5.1 Definition of Matching Network Loss

Loss definitions are varied. For RF and microwave network analysis, return loss and transmission loss are common since they are derived on the basis of traveling waves and interfacial scattering. The matching network loss definition most meaningful to assess the impact on power and PAE is the loss that results under simultaneous conjugate match at both ports, as this implies all available source power is delivered to the matching network load, any deviation from this ideal due strictly to heat, which is unrecoverable.[6] This definition is due to heating loss exclusively, though it is often referred to simply as insertion loss (IL). The power delivered to the load is always less than available power from the transistor since no physical network is lossless.

Expressed by s-parameters as

$$IL = \frac{|S_{21}|^2}{1 - |S_{11}|^2} \tag{6.9}$$

---

6 Radiation effects are ignored, but can be substantial at mm-wave frequencies. Radiation would appear as loss in a rigorous accounting.

it is important to understand in general that IL does not equal $|S_{21}|^2$. Figure 2.2 shows the two-port definitions associated with Eq. (6.9), with the standard convention of port one to port two going from left to right, meaning that port two would drive the transistor and port one would load it. It is interesting to note by invoking energy conservation that, for a lossless network, $|S_{11}|^2 + |S_{21}|^2 = 1$, which upon substitution into Eq. (6.9) yields 0 dB insertion loss, as expected.

6.5.2 The Effects of Matching Network Loss

Considerable effort is allocated to maximize the PAE of the PA, including use of load-pull for identification of optimum harmonic terminations and advanced transistor material technology and structures. It is surprising, however, how little attention is placed on optimizing insertion loss of the matching network. For example, the use of a low-cost chip capacitor might yield an insertion loss of 1 dB, resulting in a PAE decrease from 70% to 55%, whereas a low-loss capacitor yields an insertion loss of 0.5 dB and an associated PAE of 62%. Understanding loss mechanisms in the RF matching network, their physical origins, and how they can be minimized is thus a central component of the many tactics in achieving the highest possible PAE.

Figure 6.17 plots apparent PAE for load matching network insertion, loss from 0 to 1.0 dB, with each curve corresponding to raw transistor efficiency at 0 dB insertion loss. For example, under the worst-case loss, for an ideal PA with PAE of 100%, the loss due to the matching network would yield a PAE of only 79%.

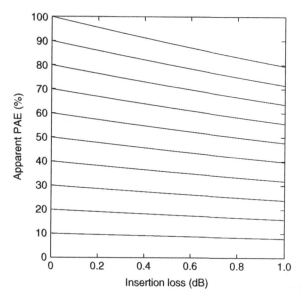

Figure 6.17 Effective PAE versus load matching network insertion loss.

As the efficiency gets higher, matching network loss becomes more apparent, and troublesome.

6.5.3 Minimizing Matching Network Loss

Figure 6.18 illustrates a generalization of the low-pass L-section network of Figure 6.2, by the addition of series resistance and shunt resistance, to model inductor loss and capacitor loss, respectively. A somewhat similar abstraction can be derived for the distributed-network line by the Telegrapher's equations, but will not be repeated presently, being well treated in [14].

In effecting impedance matching, circulating current in the shunt capacitor arm of Figure 6.18, denoted i_C, can be substantial, as the capacitor reduces the magnitude of the impedance R_H in parallel with C, with inductor L advancing the phase to present a pure resistance at resonance. Because current i_C is proportional to Q_0, matching network insertion loss due to shunt capacitance is directly proportional to Q_0^2. To generate the most common load-lines found in the wireless PA, the impedance transformation ratio generally spans 5–10. It is evident that for a fixed impedance transformation ratio in this range that a multi-section matching network will exhibit lower insertion loss than a single-section matching network because circulating current is substantially reduced by substituting a high-Q single-section matching network by a low-Q multi-section matching network.

That additional sections yield lower insertion loss appears counter-intuitive. For the present application, this phenomena occurs because the parasitic resistances R_C and R_L are approximately equal, while the shunt current i_C is a magnitude higher than series current i_L. Three significant consequences result. The number of sections, N, is now subject to IL as an additional optimization constraint. A minimum insertion loss may impose a larger number of sections,

Figure 6.18
Generalization of
L-section to include series
and shunt loss elements.

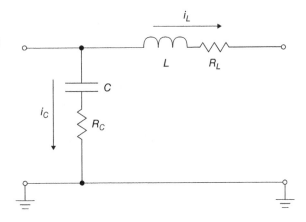

N, than prescribed by Eq. (6.3), which is based exclusively on the impedance transformation ratio.

The optimum location of series capacitance, such as the DC blocking capacitor, can be deduced by observing the relative maxima and minima of voltage and current at each node of the matching network. Similarly, the optimum location of the bias feed, whose impedance must be substantially larger than the impedance at the point it loads, can be identified, to minimize parasitic loading and maximize isolation between the RF matching network, the transistor, and bias network. This property is why the bias feed is often placed near the transistor, at a low impedance node, and the DC blocking capacitor is usually placed at a high impedance node where its equivalent series resistance (ESR) is negligible with respect to 50 Ω and current is at a relative minimum. Optimum bias feed isolation also simplifies second-harmonic loading with the bias network.

6.6 Optimum Harmonic Termination Design

This section presents an abbreviated review of optimum harmonic termination theory followed by comprehensive treatment of optimum harmonic termination synthesis. It is assumed the reader is with optimum harmonic termination theory as found in [15, 16].

6.6.1 Optimally Engineered Waveforms

A significant consequence of large-signal operation of a transistor, defined as the load-line trajectory traversing from maximum current to maximum voltage, is substantial harmonic generation. Systematic and deliberate exploitation of this phenomena is commonly referred to as waveform engineering, with well-established theoretical treatment in [15–18]. Presently it assumed the reader is adequately acquainted with these references to permit the present treatment to focus exclusively on design methods to exploit harmonic generation for efficiency and linearity enhancement.

Optimum harmonic termination design, easily implemented by frequency-domain methods, follows from Fourier decomposition of a periodic signal to emulate a waveform exhibiting properties amenable to a desirable property, such as high efficiency. Depending on the ratio of the PA operating frequency to the f_T of the transistor, substantial improvements in efficiency can be realized by optimum second-harmonic termination, and to a lesser degree, optimum third-harmonic termination. Similarly, substantial improvements in linearity can be realized by optimum envelope termination, a second-order mixing product, as it influences in-band linearity [19]. A simple illustration of the underlying principle of harmonic termination optimization is made by considering the first three terms of the drain-source (collector-emitter) voltage of a transistor operating in compression

$$v(t)_{DS} \approx V_{DSQ} + V_1 \sin(\omega\, t) + V_2 \sin(2\omega\, t + \phi_2) + V_3 \sin(3\omega\, t + \phi_3) \quad (6.10)$$

where V_2, ϕ_2, V_3, and ϕ_3 are established by load impedance at ω_2 and ω_3. In recalling the operation of an ideal switch, it is reasonable to conjecture that elimination of even-order harmonics and optimum amplitude and phase of odd-harmonics would emulate such a switching waveform. Following what will be defined as the Raab criteria to establish maximum efficiency there results

$$v(t)_{DS} \approx V_{DSQ} + V_1 \sin(\omega\, t) + 0.125V_1 \sin(3\omega\, t) \quad (6.11)$$

with $V_3 = 0.125V_1$ exhibiting a maximally flat drain-source voltage, yielding maximum efficiency [20, 21]. An illustration of this voltage is shown in Figure 6.19 where it seen that in-phase addition of an optimum third-order contribution flattens the drain-source voltage to emulate an ideal switch. In practice, finite drain-source saturation voltage and a nonzero second-harmonic contribution will cause a change from the ideal presented here, which nevertheless delivers a theoretical efficiency of slightly over 88%, and in practice can achieve over 85% with gallium arsenide (GaAs) and gallium nitride (GaN) technologies and over 75% for LDMOS technology.

Figure 6.19 An optimally engineered drain-source voltage waveform, satisfying the Raab criteria for maximum efficiency, composed of a zero second-harmonic component and an in-phase third-harmonic component of $0.125V_1$. The DC quiescent voltage has been removed.

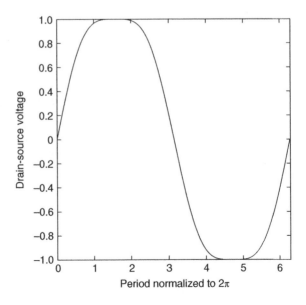

6.6.2 Physical Implementation of Optimum Harmonic Terminations

Optimally engineered waveforms resembling those Figure 6.19 are typically associated with Class F and Class J modes, whose underlying theoretical foundations are built on frequency-domain physical implementation criteria.[7] In the Class F forward mode, the drain-source voltage is loaded by an equivalent infinite cascade of series and parallel resonators exhibiting zero impedance at even harmonics and infinite impedance at odd harmonics, respectively, replicating a switching waveform in the time-domain. In the frequency-domain, this load is easily implemented by synthesizing a short-circuit at even harmonics and an open-circuit at odd harmonics, with these conditions defined precisely at the reference-plane defining the active area of the transistor, as illustrated in Figure 6.20.

A common physical implementation of optimum harmonic terminations is easily achieved by exploiting the isolation provided by the matching bias network's periodic response over harmonics, which effectively decouples the load impedance at the fundamental and harmonic impedance terminations. Decoupling also enables the ability to independently tune the impedance at the fundamental and harmonics, thereby providing an invaluable capability in empirical optimization of final performance. Alternatively, in those applications where a distributed-network feed cannot be used for bias, a lumped network can be added in parallel to the load network to synthesize an optimum second harmonic impedance consisting of a wire-bond and a shunt capacitor to ground resonant at the second harmonic. Isolation from this network and the fundamental relies on high-Q, though some iteration may be necessary.

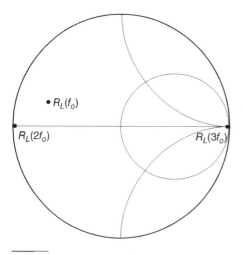

Figure 6.20 Transistor load terminations at the fundamental and harmonics for optimum efficiency normally associated with Class F operation. For most wireless applications, harmonic termination above third-order is seldom necessary. The Smith chart reference impedance is 50 Ω.

---

[7] In contrast, the underlying theory of Class E rests on a time-domain foundation by establishing specific criteria for the instantaneous drain-source voltage and its first derivative.

In exploiting the periodic response of the bias network, which presents an open circuit at the fundamental, the drain sees the fundamental load impedance provided by the matching network and a short circuit at the second harmonic provided by the bias network. The bias network feed consists of a quarter-wave line terminated by a shunt capacitor to ground that is chosen to be series resonant somewhere between the fundamental and second harmonic. In practice, isolation is not complete, and mutual interaction exists between the load impedance provided by the matching network, the second harmonic of the matching network, and the second harmonic impedance of the bias network, thus requiring iteration with both simulation tools and with the physical implementation. This method can be extended to third harmonic terminations using a similar approach, one common method being a $\lambda/12$ stub.

6.6.3 Optimum Harmonic Terminations in Practice

The literature on optimum harmonic impedance termination design identifies results that in practice are difficult, if not impossible, to achieve. No matching network is entirely decoupled between fundamental and harmonics, though this itself is seldom a limiting factor. Parasitic loss often represents the limiting factor in synthesis of optimum harmonic terminations, which in practice will be bounded by a maximum gamma magnitude of length 0.95–0.98, depending on frequency, substrate quality, and passive component quality. Similarly, passive load-pull systems are in practice limited to harmonic impedance boundaries not much different from that obtainable with a physical network, thereby providing a degree of realism in that load-pull provides an impedance that is physically realizable. From this observation follows three significant corollaries next considered:

Optimum harmonic termination impedance depends on the relative ratio to the fundamental impedance, not the absolute value: Equation (6.10) illustrates that the relative voltage contribution of each spectral component establishes the optimum voltage of Eq. (6.11). Therefore, the harmonic impedances need only be large enough to satisfy the Raab criteria relative to the fundamental for an optimally engineered Class F drain voltage. In practice, this can be achieved with a ratio of 10× between fundamental and harmonic termination impedances. This corollary substantially relaxes the conditions often cited in the theoretical literature that optimum harmonic terminations be a perfect short or open. Similarly, in load-pull, it follows that harmonic termination synthesis need not realize unit reflection coefficient magnitude, contrary to common belief.

Matching network insertion loss has a large and direct influence on PAE: Though substantial effort is expended in extracting as much efficiency from the PA as possible, use of low-quality passive components, as reflected by relatively larger capacitor and inductor ESR or higher substrate loss-tangent, can reduce, or eliminate, any gains achieved by load-pull optimization or harmonic termination optimization. The differential expense added by high-performance matching network components is far lower than trying to achieve high efficiency through esoteric architectures and expensive transistor technologies.

Harmonic termination maxima are generally smooth and relatively flat near optimum: Successful harmonic termination does not require a precise short or open. Except in rare occasions, locating the harmonic termination within 10° of the optimum identified by load-pull will yield adequate improvement while simultaneously enhancing process, voltage, and temperature (PVT) robustness.

6.7 Closing Remarks

The central feature of the load-pull method of RF PA design is its distinction between management of nonlinearity versus analysis of nonlinearity, transforming what is essentially an intractable nonlinear mathematical problem into a series of variable-impedance measurements followed by utilization of well-understood linear network theory for matching network design. Following identification of optimum terminating impedances by load-pull, at all relevant mixing products, the matching network replicates these conditions using well-defined and structured linear synthesis and physical implementation methods. What if a complex nonlinear problem becomes, then, principally a problem of matching network design.

The network synthesis methods treated in this chapter exhibit specific properties ideal for RF power amplification, including impedance transformation ratio, bandwidth, insertion loss, harmonic rejection, and physical size. Secondary factors include out-of-band response for stability control and VBW enhancement BOM cost. The Butterworth synthesis is the principal response, based on its optimally flat bandwidth for a given number of sections, and the relative ease of solution based on graphical methods. Of the three physical implementation methods introduced, hybrid-parameter was shown to be the most useful and flexible, offering a reasonable compromise in performance and physical size and ease of post-fabrication tuning for virtually all PA applications, spanning 2G to 4G and mobile to base-station.

Of the three additional synthesis methods introduced as compliments to the Butterworth response, the Hecken response was identified as providing the maximum simultaneous discontinuity-free impedance transformation and bandwidth for a specified electrical length. The Hecken taper was introduced primarily for load-pull test-fixture applications where expanded bandwidth is necessary for multi-harmonic load-pull and instantaneous frequency agility that are impossible with individual quarter-wave sections or difficult with multi-section transformers.

The theory and practice of optimum harmonic terminations were explored. Starting with a pedagogical time-domain treatment of Class F, the Raab criteria was introduced to identify the optimal third-harmonic termination for a maximally flat voltage, yielding maximum PAE. Class F in the frequency-domain was next treated, including introduction matching network augmentation by practical harmonic termination implementation.

References

1 Guillemin, E.A. (1957). *Synthesis of Passive Networks*. New York: Wiley.

2 Guillemin, E.A. (1935). *Communication Networks*. New York: Wiley.

3 Valkenburg, M.E.V. (1960). *Introduction to Modern Network Synthesis*. New York: Wiley.

4 Bode, H.W. (1945). *Network Analysis and Feedback Amplifier Design*. New York: Van Nostrand.

5 Tuttle, D.F. (1958). *Network Synthesis*. New York: Wiley.

6 Matthaei, G.L., Young, L., and Jones, E.M.T. (1980). *Microwave Filters, Impedance-Matching Networks, and Coupling Structures*. London: Artech House.

7 Cristal, E.G. (1965). Tables of maximally flat impedance-transforming networks of low-pass-filter form. IEEE Transactions on Microwave Theory and Techniques.

8 Matthaei, G.L. (1964). Tables of Chebyshev impedance-transforming networks of low-pass filter form. IEEE Transactions on Microwave Theory and Techniques.

9 Hecken, R.P. (1972). A near-optimum matching section without discontinuities. IEEE Transactions on Microwave Theory and Techniques.

10 Collin, R.E. (1956). The optimum tapered transmission line matching section. *Proceedings of the IRE* 44: 539–548.

11 Walker, L.R. and Wax, N. (1945). Nonuniform transmission-lines and reflection coefficients. *Journal of Applied Physics* 17: 1043–1045.

12 Klopfenstein, R.W. (1955). A transmission line design of improved taper. Proceedings of the IRE.

13 Thomson, W.E. (1952). Networks with maximally flat delay. *Wireless Engineer* 29: 255–263.

14 Collin, R.E. (1967). *Foundations of Microwave Engineering*. New York: McGraw-Hill.

15 Cripps, S.C. (2002). *RF Power Amplifiers for Wireless Communications*, 2e. London: Artech House.

16 Cripps, S.C. (2008). *Advanced Techniques in RF Power Amplifier Design*. London: Artech House.

17 Bostian, C.W. and Raab, F.H. (1980). *Solid-State Radio Engineering*. New York: McGraw-Hill.

18 Maas, S.A. (1988). *Analysis of Nonlinear Microwave Circuits*. London: Artech House.

19 Sevic, J.F. and Steer, M.B. (1998). A novel envelope termination method for ACPR optimization of RF and microwave power amplifiers. IEEE Transactions on Microwave Theory and Techniques Symposium Digest.

20 Raab, F.H. (1997). Class-F power amplifiers with maximally flat waveforms. IEEE Transactions on Microwave Theory and Techniques Symposium.

21 Raab, F.H. (2001). Maximum efficiency and output of class-F power amplifiers. IEEE Transactions on Microwave Theory and Techniques Symposium.

Index

The Load-Pull Method of RF and Microwave Power Amplifier Design, First Edition. John F. Sevic.
© 2020 John Wiley & Sons, Inc. Published 2020 by John Wiley & Sons, Inc.